Haruto

Akari

Sara

Yu

JN119582

2nd Grade Vol. 2

Study with your Friends!

How do we learn mathematics?

Find the ?
So many tulips bloomed!
There are 13 yellow ones.

Based on the problem you find in your daily life or what you have learned, let's come up with a purpose.

The first problem of the lesson is written. On the left side, what you are going to learn from now on through the problem is written.

When you see the problem and think that you "want to think", "want to represent", "want to know", and "want to explore", that will be your "purpose" of your learning. You can find the purpose not only at the beginning of the lesson but in various timings and settings.

You can check your understanding and try more using what you have learned.

①

Let's try this problem first.

The starting point

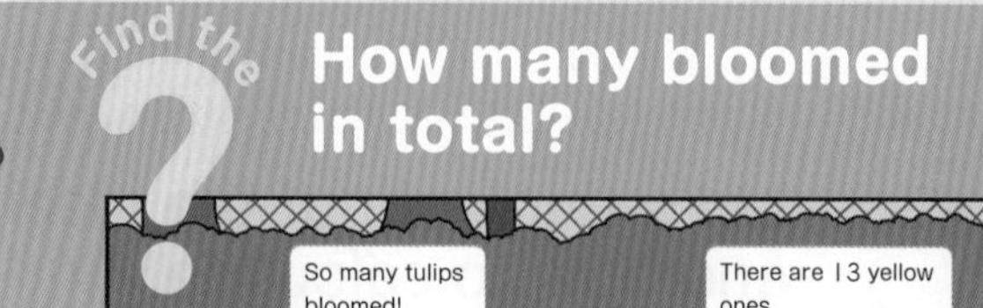

What you have learned today

Addition in Vertical Form

Let's think about the meaning of addition and how to add.

1 Addition of 2-digit numbers

Addition in vertical form

1 Tulips have bloomed. There are 24 red tulips and 13 yellow tulips.
How many tulips are there in total?

Want to know

? Purpose How do we add in vertical form?

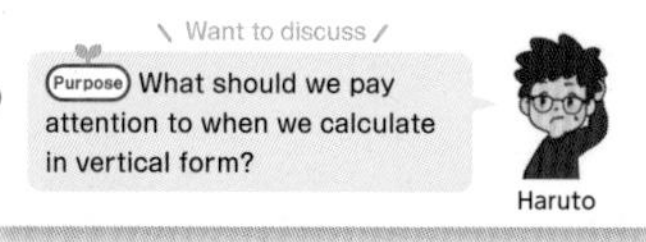

Want to discuss
Purpose What should we pay attention to when we calculate in vertical form?
Haruto

1 Let's calculate the following in vertical form.

①

	3	1
+	5	7

②

	1	7
+	3	0

③ 15 + 62
④ 65 + 31
⑤ 18 + 40

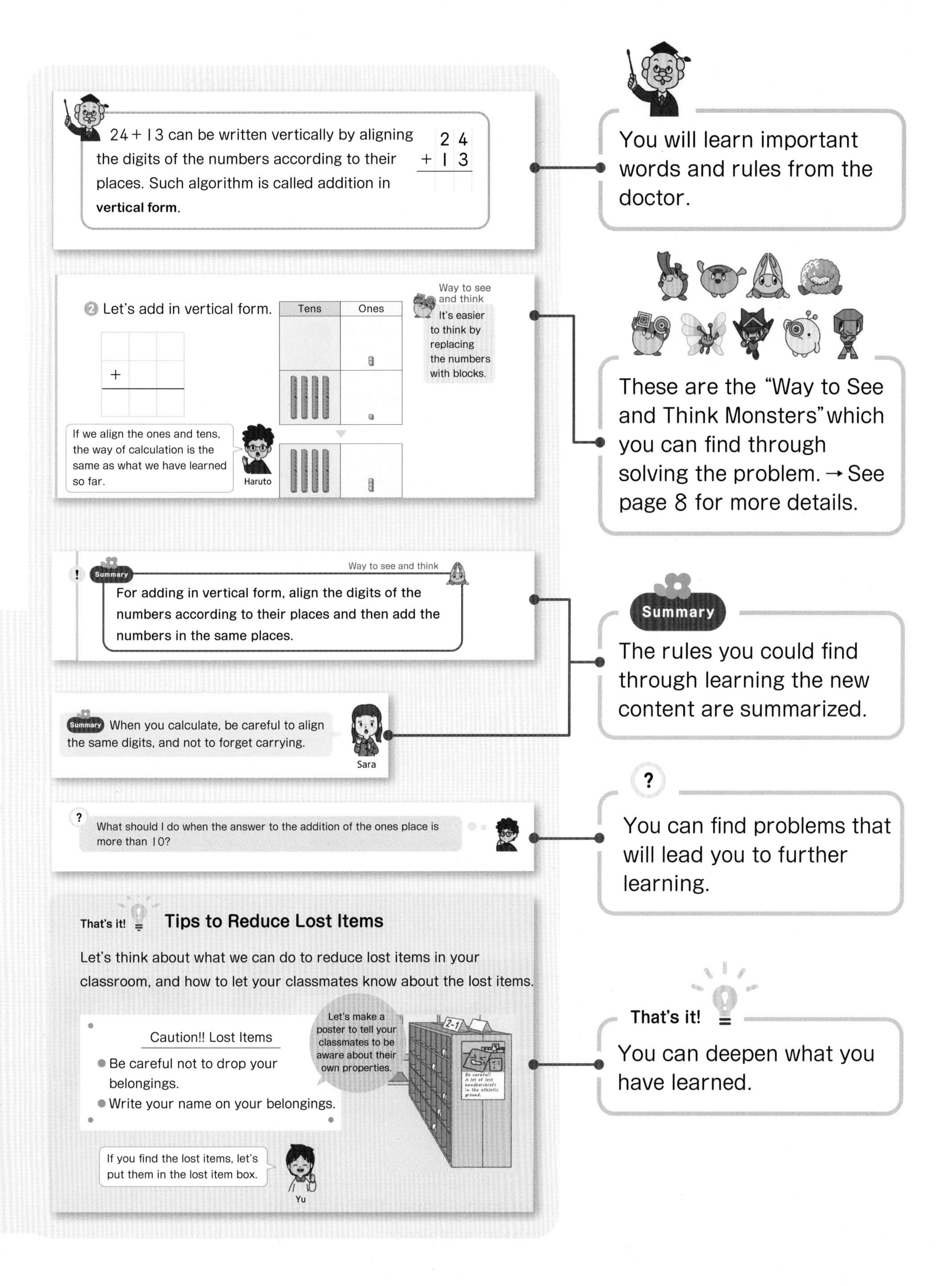
24 + 13 can be written vertically by aligning the digits of the numbers according to their places. Such algorithm is called addition in vertical form.
2 4
+ 1 3
You will learn important words and rules from the doctor.
② Let's add in vertical form.
+
Tens
Ones
Way to see and think
It's easier to think by replacing the numbers with blocks.
If we align the ones and tens, the way of calculation is the same as what we have learned so far.
Haruto
These are the "Way to See and Think Monsters" which you can find through solving the problem. → See page 8 for more details.
Summary
Way to see and think
For adding in vertical form, align the digits of the numbers according to their places and then add the numbers in the same places.
Summary
The rules you could find through learning the new content are summarized.
Summary
When you calculate, be careful to align the same digits, and not to forget carrying.
Sara
?
What should I do when the answer to the addition of the ones place is more than 10?
?
You can find problems that will lead you to further learning.
That's it!
Tips to Reduce Lost Items
Let's think about what we can do to reduce lost items in your classroom, and how to let your classmates know about the lost items.
Caution!! Lost Items
Be careful not to drop your belongings.
Write your name on your belongings.
Let's make a poster to tell your classmates to be aware about their own properties.
2-1
If you find the lost items, let's put them in the lost item box.
Yu
That's it!
You can deepen what you have learned.

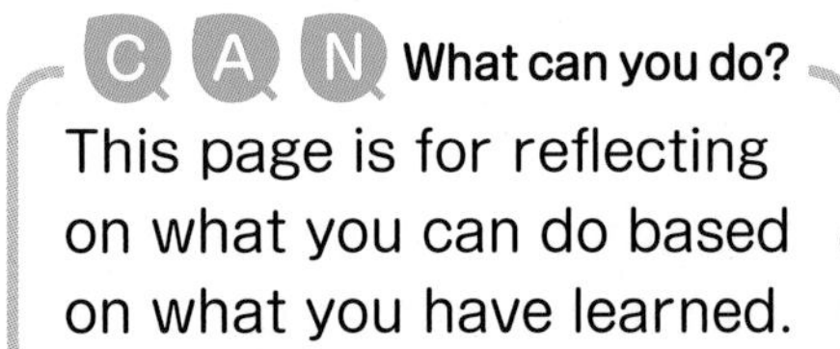

This page is for reflecting on what you can do based on what you have learned.

You will talk about which "Way to See and Think Monsters" you found in the process of learning.

Utilize Usefulness and Efficiency of Learning

This page is for trying to solve a wide variety of problems based on what you have learned.

With the Way to see and think Monsters...

Let's Reflect!

This page is for reflecting on what you have learned with the "Way to See and Think Monsters."

Solve the ? Want to Connect

This page is for solving problems based on what you have learned. Moreover, it is for trying to find the next "?" that connects to further learning.

This page is for reflecting on what you have learned, and connecting them to your further learning.

Summarizing and Reflecting what you have learned

CAN What can you do?

We understand the meaning of the way to add in vertical form. → p.43

1 Let's summarize how to add $67+28$ in vertical form.

① In the ones place, $7+8$ makes 15.

Which "Way to See and Think Monsters" did you find in "4 Addition in Vertical Form"?

Yu: I found "Align" when I was calculating addition in vertical form.

Akari: I found other monsters too!

Utilize Usefulness and Efficiency of Learning

1 Find the mistakes in the following processes.

Let's write the correct answer in the ( ).

With the Way to see and think Monsters...

Let's Reflect!

Let's reflect on which monster you found in "4 Addition in Vertical Form."

Align

For addition in vertical form, we could calculate by aligning the digits of the numbers according to their places and then

Solve the ?

For adding 2-digit numbers in vertical form, we can calculate by adding the numbers in the ones place and then the numbers in the tens place.

Haruto

Want to Connect

Can we calculate in the same way even when the numbers get larger?

Akari

51

Reflect

Connect

lem Let's compare addition and subtraction in vertical form.

s calculate $59+73$ in vertical form.

This page is for reviewing areas where you have difficulties to solve the problem, or are likely to make mistakes.

Utilizing Math for SDGs

Summer is coming!
How can we spend a happy and healthy summer vacation?

SUSTAINABLE DEVELOPMENT GOALS

Utilizing Math for SDGs

The Sustainable Development Goals (SDGs) are a set of goals that we aim to achieve in order to create a world where we can live a life of safety and security. This page will help you think about what you can do for society and the world through math.

About the QR Code

Some of the pages include the QR code which is shown on the right.

▷ … You can learn how to draw a diagram and how to calculate by watching a movie.

… You can learn by actually moving and operating the contents.

… You can learn by reflecting on what you have learned previously in your previous grades.

… You can utilize it to know the solution to the problems that you couldn't find out the answer, or to try various problems.

… You can deepen your learning by actually looking at the materials including the website.

Dear Teachers and Parents

This textbook has been compiled in the hope that children will enjoy learning through acquiring mathematical knowledge and skills.
The unit pages are carefully written to ensure that students can understand the content they are expected to master at that grade level.
In addition, the "More Math!" section at the end of the book is designed to ensure that each student has mastered the content of the main text, and is intended to be handled selectively according to the actual conditions and interests of each child.
We hope that this textbook will help children develop an interest in mathematics and become more motivated to learn.

The sections marked with this symbol deal with content that is not presented in the Courses of Study for that grade level, thus does not have to be studied uniformly by all children.

QR codes are used to connect to Internet content by launching a QR code-reading application on a smartphone or tablet and reading the code with a camera.
The QR Code can be used to access content on the Internet.
The code can also be used at the address below.

https://r6.gakuto-plus.jp/s2a0 l

Note: This book is an English translation of a Japanese mathematics textbook. The only language used in the contents on the Internet is Japanese.

【Infectious Disease Control】
In this textbook, pictures of activities and illustrations of characters do not show children wearing masks, etc., in order to cultivate children's rich spirit of communicating and learning from each other. Please be careful to avoid infectious diseases when conducting classes.

Becoming a Writing Master

The notebook can be used effectively.

- To organize your own thoughts and ideas
- To summarize what you have learned in class
- To reflect on what you have learned previously

Let's all try to become notebook masters.

Write today's date. → October 21st

Write the problem of the day that you must solve. → Problem

How many eggs are there in total?
Let's think about how to calculate.

Math Expression : 3 + 9

Let's write down what you thought while thinking about the solution of the problem as "purpose." → (Purpose)
which number should I make it 10 when I calculate?

Write your ideas or what you found about the problem. → ○ My idea
To make 10, I need 3 and 7 more.
Divide 9 into 7 and 2.
3 and 7 make 10.
10 and 2 make 12.

Tips for Writing 1

Answer: 12 eggs ← Don't forget to write "eggs" here!

Tips for Writing ❶

When you made a mistake, don't erase it so that it will be easier to understand when you look back at your notebook later.

Tips for Writing ❷

By finding the "Way to See and Think Monsters," it will connect you to what you have learned previously.

Tips for Writing ❸

By writing down what you would like to try more, it will lead you to further learning.

○ Yu's idea

To make 10, I need 9 and 1 more.
Divide 3 into 1 and 2.
9 and 1 make 10.
10 and 2 make 12.

Divide

where did I find this before?

Tips for Writing ❷

Answer: 12 eggs

It is easier to make 10 by dividing.

⟨Summary⟩

The answer is the same, whichever number is made to be 10 and calculated.

⟨Reflection⟩

I found that I could do it whichever way was easier for me since it is okay to make 10 in either way.

⟨What I want to do next⟩

I want to do addition of larger numbers.

Write the classmate's ideas you consider good.

Summarize what you have learned today.

Reflect on your class, and write down the following;

- What you learned.
- What you found out.
- What you can do now.
- What you don't know yet.

While learning mathematics...

Based on what I have learned previously...

Why does this happen?

There seems to be a rule.

You may be in situations like above. In such case, let's try to find the "Way to See and Think Monsters" on page 9. The monsters found there will help you solve the mathematics problems. By learning together with your friend and by finding more "Way to See and Think Monsters," you can enjoy learning and deepening mathematics.

What can we do at these situations?

I think I can use 2 different monsters at the same time...

→ You may find 2 or 3 monsters at the same time.

I came up with the way of thinking which I can't find on page 9.

→ There may be other monsters than the monsters on page 9. Let's find some new monsters by yourselves.

Now let's open to page 9 and reflect on the monsters you found in 1st grade. They surely will help your mathematics learning in 2nd grade.

Representing ways of thinking in mathematics

Way to See and Think Monsters

Unit

If you set the unit...

Once you have decided one unit, you can represent how many using the unit.

Summarize

If you try to summarize...

It makes it easier to understand if you summarize the numbers.

Other Way

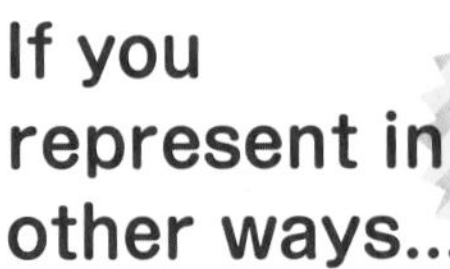

If you represent in other ways...

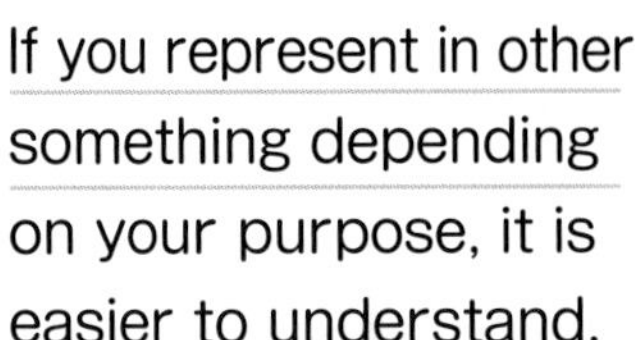

If you represent in other something depending on your purpose, it is easier to understand.

Align

If you try to align...

You can compare if you align the number place.

Change

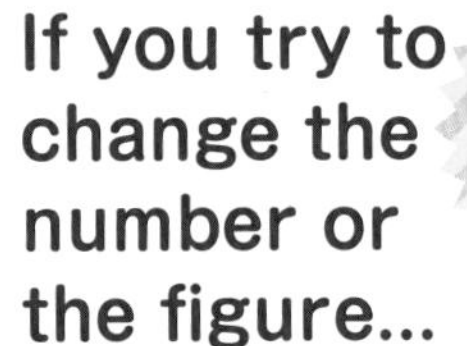

If you try to change the number or the figure...

If you try to change the problem a little, you can understand the problem better or find a new problem.

Divide

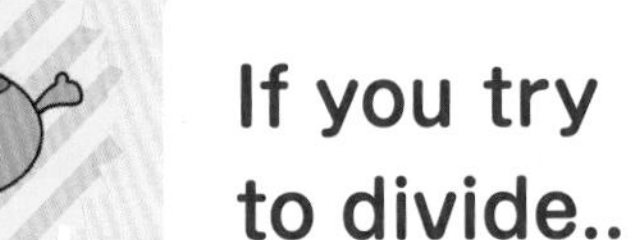

If you try to divide...

Decomposing numbers by place value and dividing figures makes it easier to think about problems.

Why

You wonder why?

Why does this happen? If you communicate the reasons in order, it will be easier to understand for others.

Rule

Is there a rule?

If you examine a few examples, you can find out whether there is a rule or not.

Same Way

Can you do it in a similar way?

If you find something the same or similar to what you have learned, you can understand.

Ways to think learned in the 1st grade

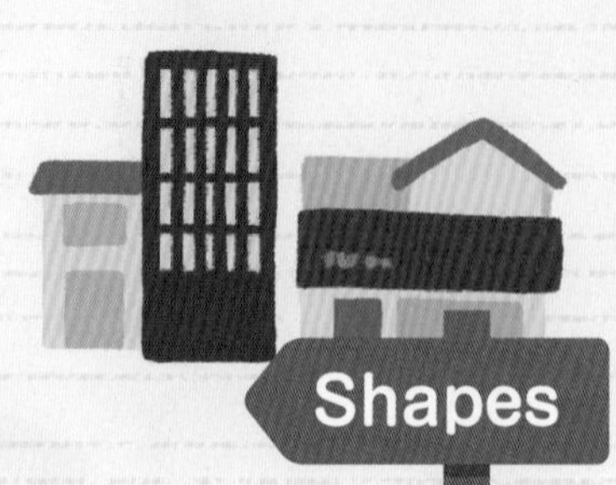

Numbers and Calculations

Divide

Dividing numbers.

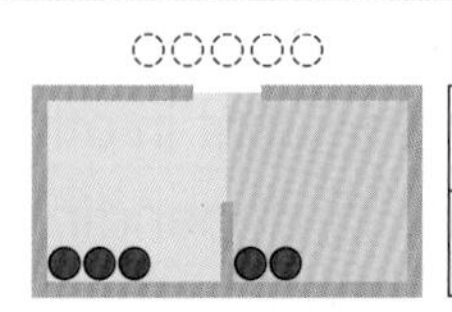

5	
3	2

Summarize

Making sets of 10.

Other way

Representing stories with pictures/ Representing stories with expressions/ Representing expressions with stories.

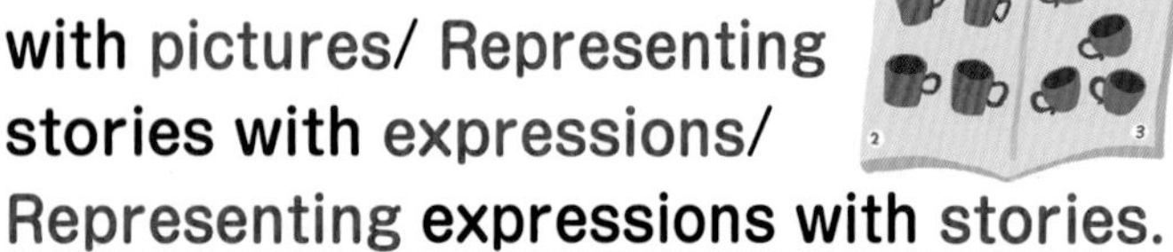

There are 6 goldfish. Put 2 more goldfish, and it becomes 8 goldfish.
Math Sentence: 6+2 = 8 Answer:  goldfish

Unit

Making sets of 10 as one unit.

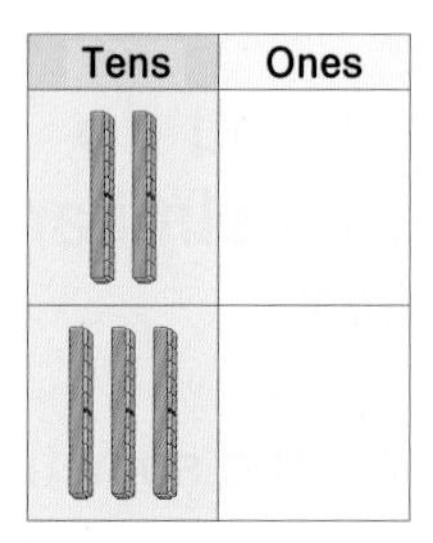

Rule

Finding out the rules by lining up the cards.

9 + 2	8 + 3	7 + 4	6 + 5	5 + 6
9 + 3	8 + 4	7 + 5	6 + 6	5 + 7
9 + 4	8 + 5	7 + 6	6 + 7	5 + 8
9 + 5	8 + 6	7 + 7	6 + 8	5 + 9

Shapes

Summarize

Classifying into the same categories.

Divide Change

Separating and connecting shapes.

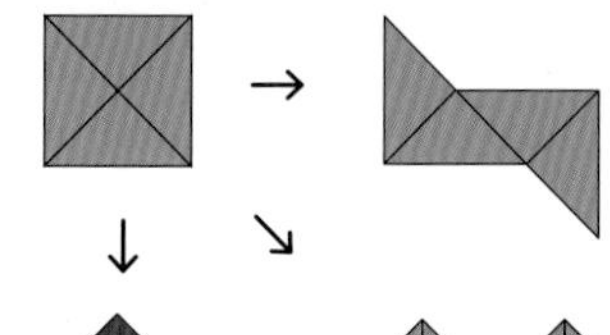

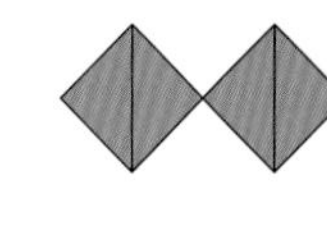

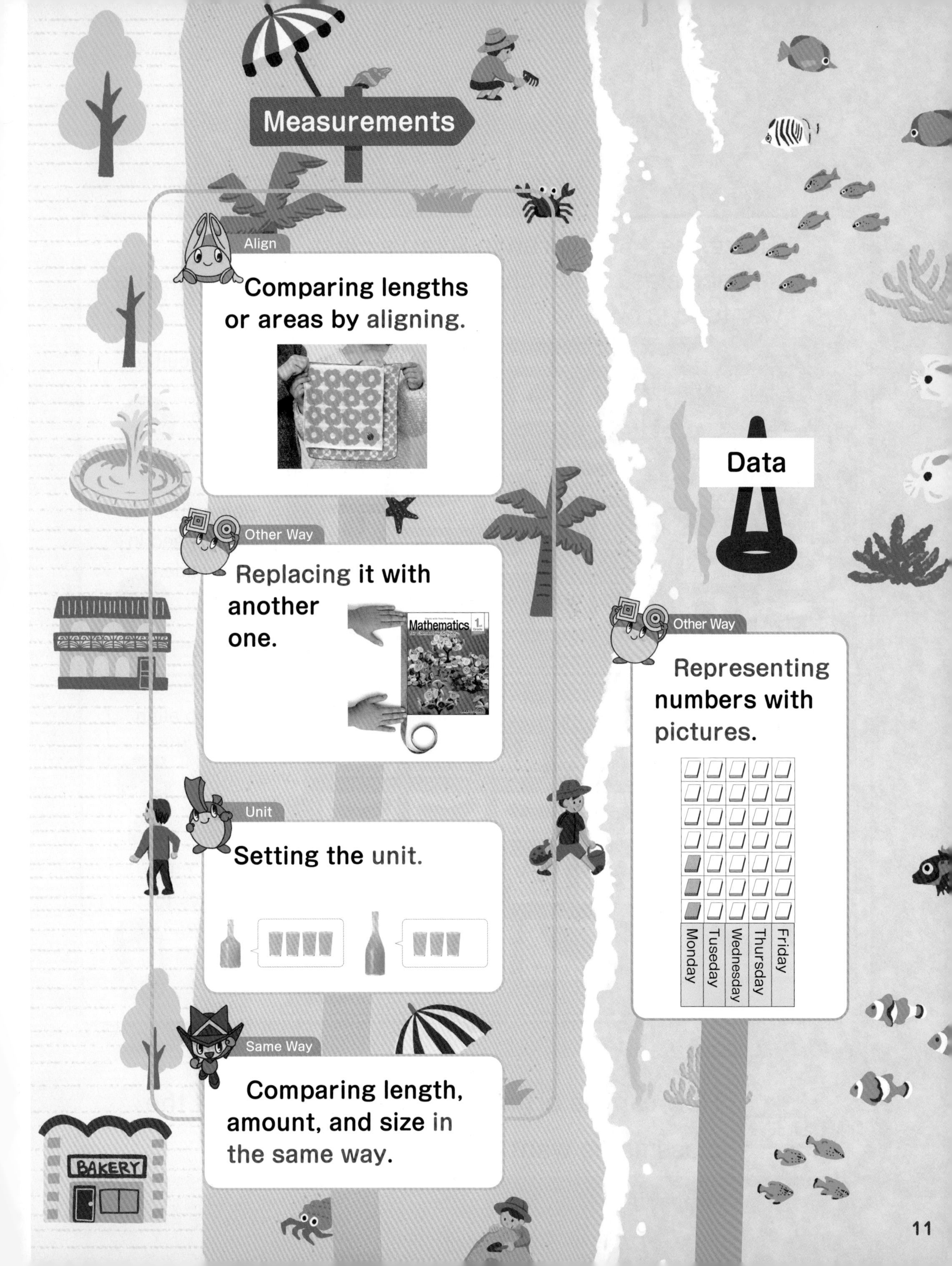
Measurements
Align
Comparing lengths or areas by aligning.
Other Way
Replacing it with another one.
Mathematics
Unit
Setting the unit.
Same Way
Comparing length, amount, and size in the same way.
BAKERY
Data
Other Way
Representing numbers with pictures.
Monday
Tuseday
Wednesday
Thursday
Friday

Find the ?

What vegetable do you want to grow?

There are 25 children in Akari's class. In the life science class, each child chose a vegetable he or she wants to grow and put on the blackboard.

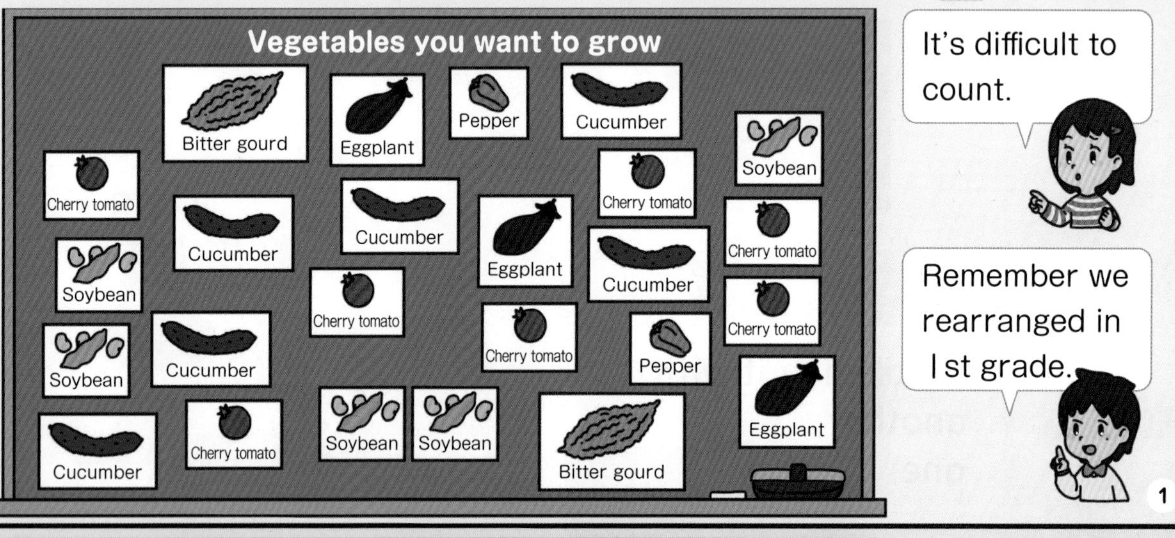

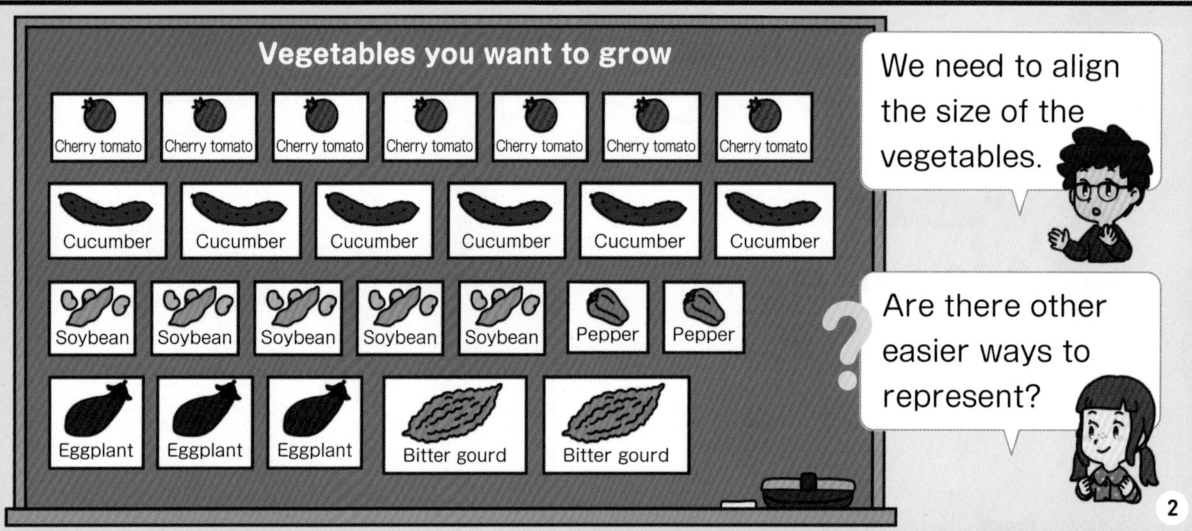

\ Want to represent /

How can we organize the vegetables that the classmates want to grow?

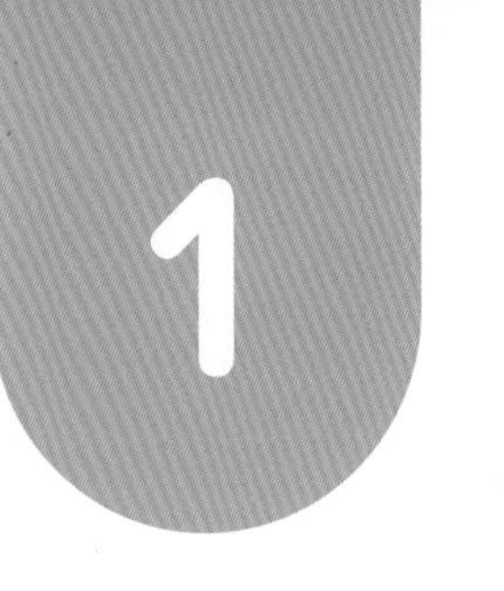

Tables and Graphs

Let's think about how to organize data and how to represent them.

How to organize ↓

1 Let's think about how to organize the data on vegetables you want to grow.

① Let's write the number of children who chose each vegetable in the **table** below.

Way to see and think
It is easier to represent on the table by sorting vegetables.

The vegetable I want to grow

Vegetables	Cherry Tomato	Cucumber	Soybean	Eggplant	Pepper	Bitter Gourd
Number of Children	7					

② Represent the number of children who chose each vegetable by using ○ on the **graph** on the right.

The vegetable I want to grow

○					
○					
○					
○					
○					
○					
○					
Cherry Tomato	Cucumber	Soybean	Eggplant	Pepper	Bitter Gourd

Summary The table is easier to read because the numbers of people who chose each vegetable are represented.

Akari

Summary The graph is better to read the difference in number.

Yu

③ What is the rank of Soybeans? How many children chose them?

Sara

Which one should I choose...

? Can we also organize other objects by using tables and graphs?

We have so many lost items recently...

What kind of lost items are found most?

2

In Haruto's class, cards shown on the right was made to find out about the lost items. On the card, the item lost and where it was lost are written. Let's find out what kind of items are lost most.

Place you lost
Lost item

Classroom	Athletic ground	Classroom	Classroom	Classroom	Athletic ground
Eraser	Handkerchief	Ruler	Pencil	Pencil	Key
Classroom	Athletic ground	Classroom	Corridor	Classroom	Athletic ground
Eraser	Pencil	Pencil	Handkerchief	Handkerchief	Handkerchief
Athletic ground	Classroom	Corridor	Classroom	Classroom	Athletic ground
Handkerchief	Pencil	Handkerchief	Textbook	Pencil	Eraser
Corridor	Classroom	Classroom	Corridor	Classroom	Classroom
Handkerchief	Key	Eraser	Key	Handkerchief	Pencil
Athletic ground	Corridor	Corridor	Classroom	Athletic ground	Classroom
Key	Eraser	Ruler	Pencil	Handkerchief	Ruler

① Let's categorize the lost items and put them in a table below.

Ways to see and think

It's easier to understand the situation using tables.

Finding out the Number of Lost Items

Lost items	Pencil	Handkerchief	Textbook	Key	Ruler	Eraser
Number of Lost Items						

② Represent the number of the lost items by using ○ on the graph on the right.

\ Want to represent /

Purpose We can see the difference easier if we use the graph, but can we find an easier way to read the graph?

We arrange the order of the items depending on what we want to focus on.

③ Let's arrange the numbers of ○ in descending order on the graph on the right below.

④ What is the most number of lost items? How many are there?

⑤ Let's talk about how to read the difference in the number of lost items.

Should I compare the numbers of ○ ?

I think it makes reading easier to write 1,2,3 ...along the horizontal line of the graph.

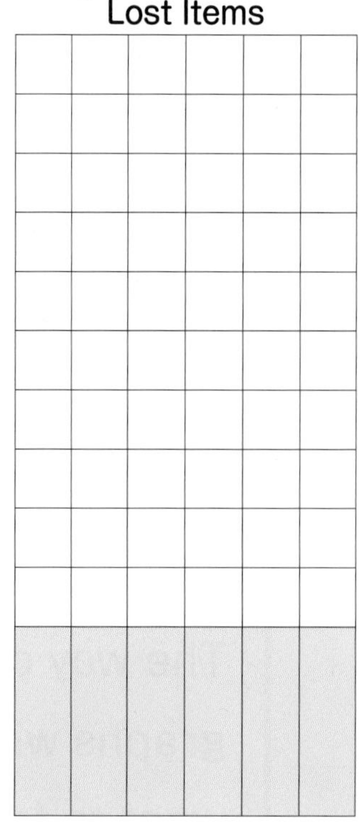

Summary It makes reading easier to arrange the numbers of ○ from the left of the graph in descending order and to write numbers along the horizontal line.

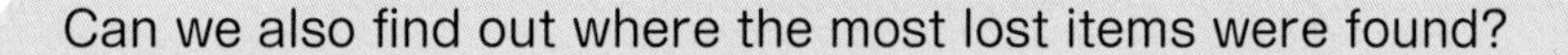

Can we also find out where the most lost items were found?

3

Where are the most number of lost items shown in 2 found? Let's find out by representing on tables and graphs.

Sara

When we wanted to find out the most number of lost items, we categorized the lost items.

Maybe we can categorize with the places this time...

Akari

? **Purpose** ＼ Want to explore ／

How will the tables and graphs change if the target changes?

The Places of the Lost Items Found

Place	Classroom	Corridor	Athletic ground
Number of Lost Items Found			

The Places of the Lost Items Found

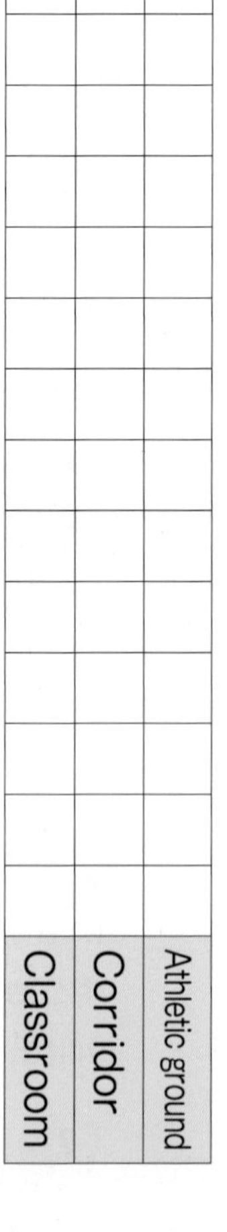

① Compared to the graph in 2, let's talk about the good points of each graph.

! **Summary**

The way of representing in tables and graphs will change depending on what you want to focus on.

1 Based on what you have found out in **2** and **3**, let's talk about what we can do to reduce the number of lost items.

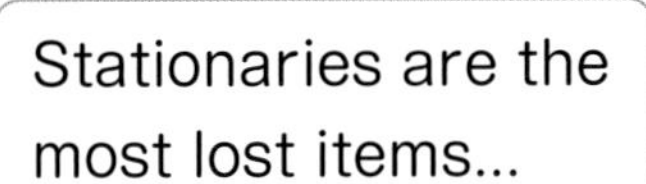

I want to be careful about losing things in the classroom...

It's better to write your name on your belongings...

How can we reduce the lost items...?

That's it! Tips to Reduce Lost Items

Let's think about what we can do to reduce lost items in your classroom, and how to let your classmates know about the lost items.

Caution!! Lost Items

- Be careful not to drop your belongings.
- Write your name on your belongings.

Let's make a poster to tell your classmates to be aware about their own properties.

2-1
Be careful!
A lot of lost handkerchiefs in the athletic ground.

If you find the lost items, let's put them in the lost item box.

Yu

CAN What can you do?

☐ We can represent what I examined on a table or a graph. → p.13

1 Let's look at the weather in March shown below.

Weather in March

1	2	3	4	5	6	7	8	9	10	11
Sunny	Sunny	Snowy	Snowy	Cloudy	Snowy	Cloudy	Rainy	Rainy	Rainy	Cloudy
12	**13**	**14**	**15**	**16**	**17**	**18**	**19**	**20**	**21**	**22**
Sunny	Sunny	Sunny	Cloudy	Cloudy	Cloudy	Sunny	Sunny	Rainy	Cloudy	Rainy
23	**24**	**25**	**26**	**27**	**28**	**29**	**30**	**31**		
Sunny	Sunny	Cloudy	Cloudy	Sunny	Rainy	Rainy	Cloudy	Sunny		

 Sunny Cloudy  Rainy  Snowy

① Write the number of days in each weather condition in the table below.

Weather in March

Weather	Sunny	Cloudy	Rainy	Snowy
Number of Days				

② Represent the numbers of days by using ○ on the graph on the right.

③ Which were more, sunny days or rainy days? By how many?

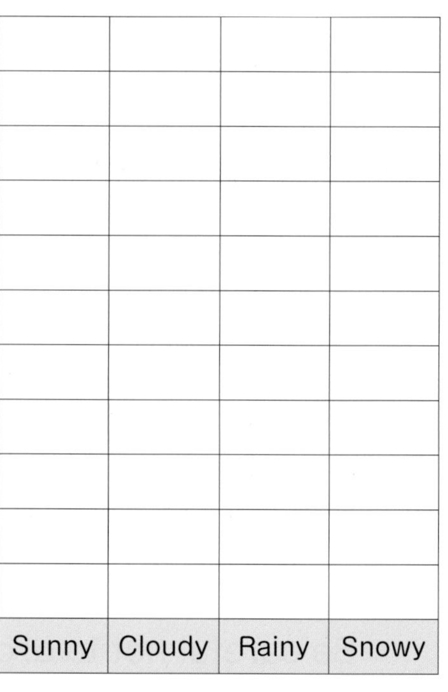

Supplementary Problems → p.151

Which "Way to See and Think Monsters" did you find in **1** Tables and Graphs"?

I found "Other Way" when I was using tables and graphs.

I found other monsters too!

With the Way to see and think Monsters...

Let's Reflect!

Let's reflect on which monster you found in "1 Tables and Graphs."

Other Way

When I represented what we want to focus on using tables and graphs, I could organize the information easily.

Summarize

The way of summarizing changed depending on what we want to focus on.

Lost items and their numbers

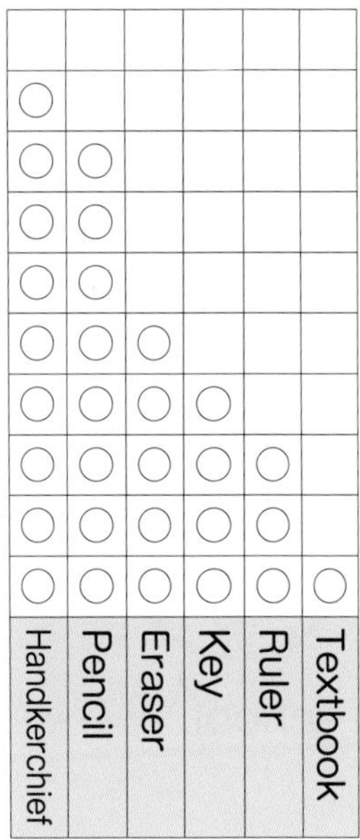

Lost items and their numbers

Classroom | Corridor | Athletic ground

Lost items and their numbers

Lost Items	Pencil	Handkerchief	Textbook	Key	Ruler	Eraser
Number of the items	8	9	1	4	3	5

It's easier to see the numbers when we use a table.

Haruto

Akari

The difference of numbers become clearer when we use a graph.

? Solve the ?

When we use tables and graphs, we can see and compare the numbers easily.

Sara

Want to Connect

What other things can we organize using tables and graphs?

Yu

Find the ?

How long were you out of school?

We went to explore our town. 

Ⓐ 

Left school.

Ⓑ 

Arrived at the fire station.

Ⓒ 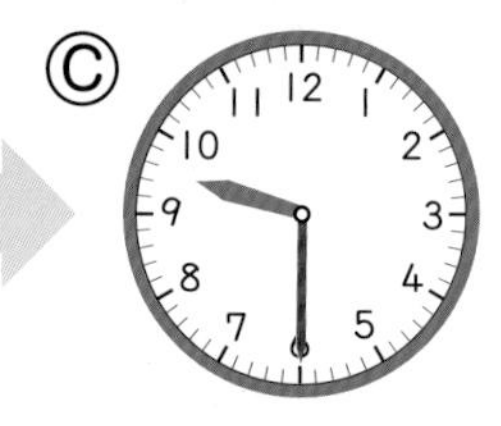

Left the fire station.

Ⓓ 

Arrived at the flower shop.

Ⓔ 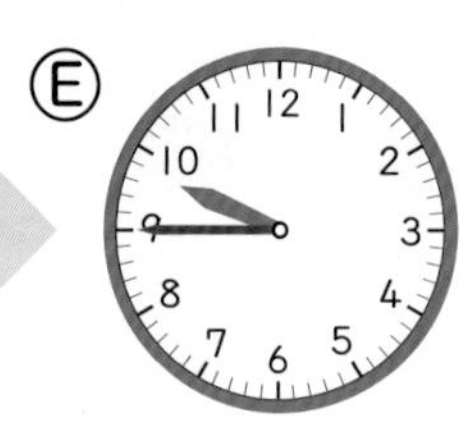

Left the flower shop.

Ⓕ 

Came back to school.

How long have you been exploring the town?

2 Time and Duration (1)

By telling time and duration, let's apply it into our life.

1 Time and Duration

How to represent time and duration →

1 Let's find out what they did in the town by telling time.

① Tell the **time** shown on each clock A, B, C, D, E, and F on page 20.

② How many marks will the long hand of the clock move from the time when they left school until the time when they arrive at the fire station?

Sara

The long hand is moving 1 mark each.

We arrived at the fire station at 9:15...

Yu

Want to explore

? **Purpose** How does the long hand of the clock move?

The duration which takes for the long hand to move from 1 scale to the next is called **1 minute**. The word minute can be written as **min**.

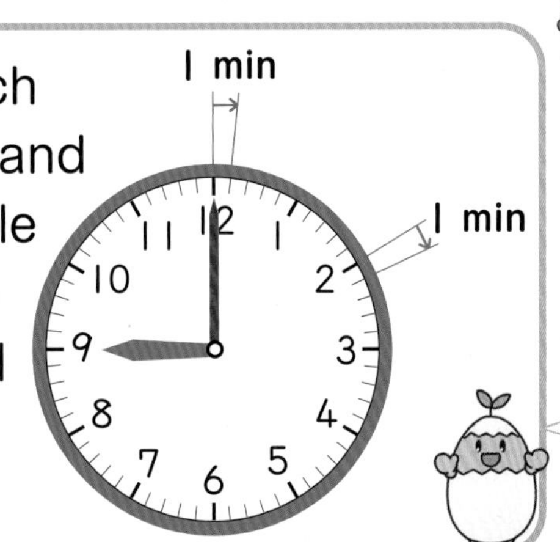

Way to see and think

We can represent various duration based on 1 min.

The time between one time and another is called duration.

If the long hand of the clock moves 15 marks, then 15 minutes have passed.

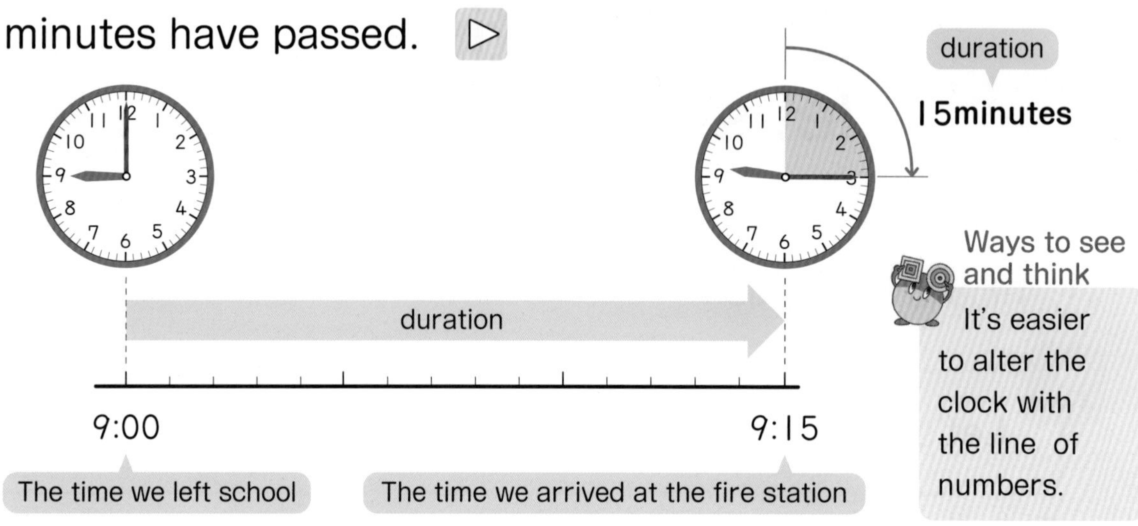

③ How much time did we spend at the flower shop?

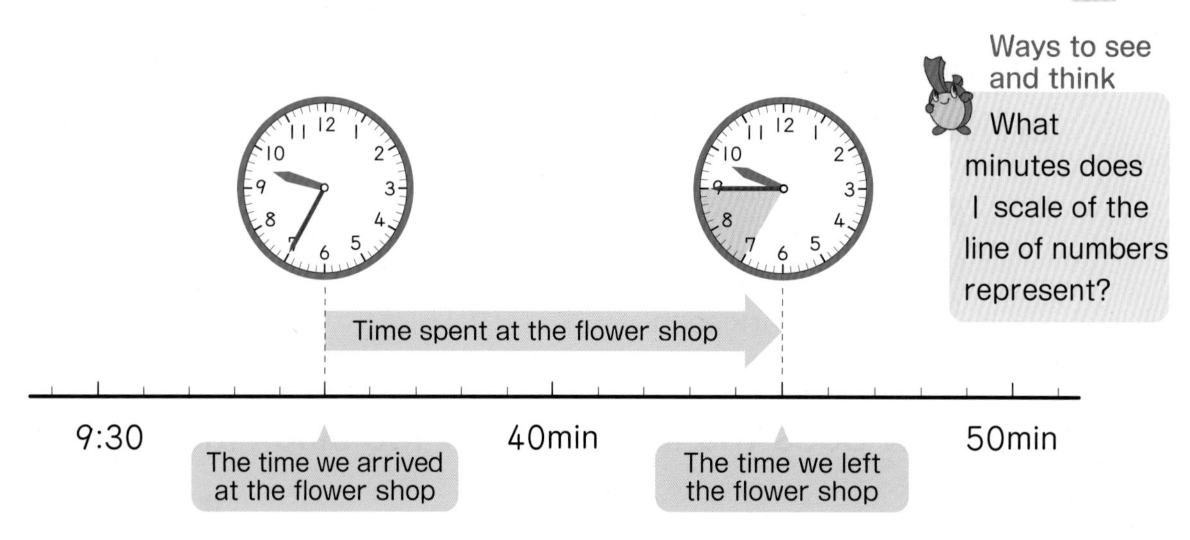

④ How long did it take from the time we left school to the time we came back to school?

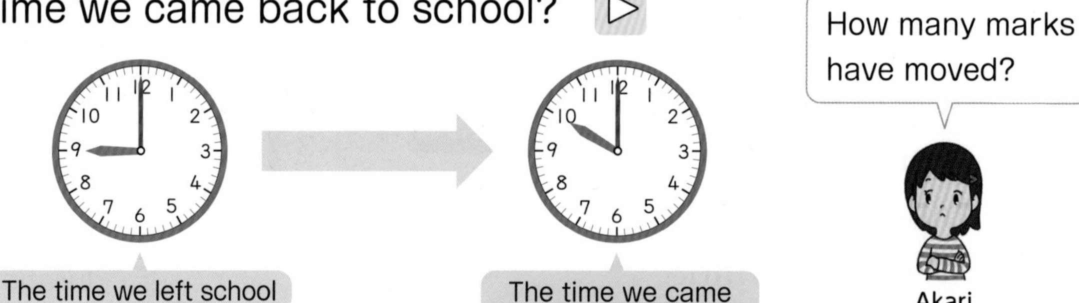

The duration which the long hand rotates completely around the clock is 60 minutes. 60 minutes is called 1 **hour**. The word hour can be written as **hr.**

1 hour = 60 minutes

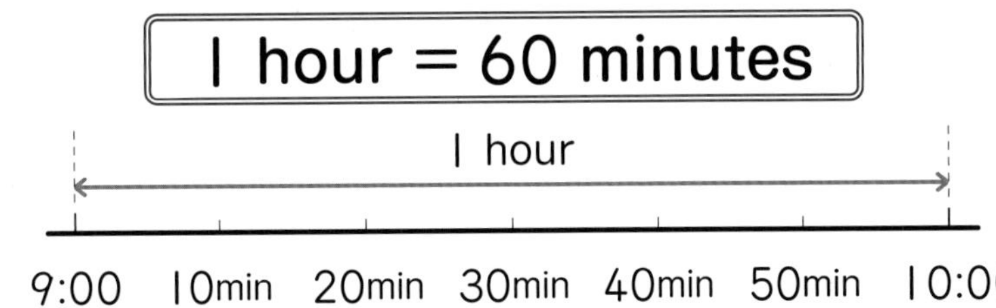

5 How long did it take from the time we left school to the time we came back to school?

It takes 1 hour for the short hand to move from 1 number to the next on the clock.

Summary

The long hand of the clock moves 1 mark in 1 minute. When it rotates completely around the clock once, that will be 1 hour.

? Can you talk about what you do in one day using time and duration?

That's it! **1 scale for the line of numbers**

Be careful on what 1 scale represents when you express clock with the number of lines.

1 scale for the line of numbers on the left side represents 1 minute.
1 scale for the line of numbers on the right side represents 10 minutes.

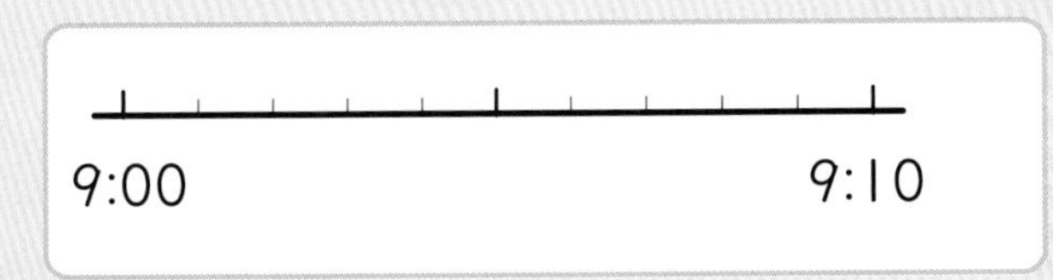

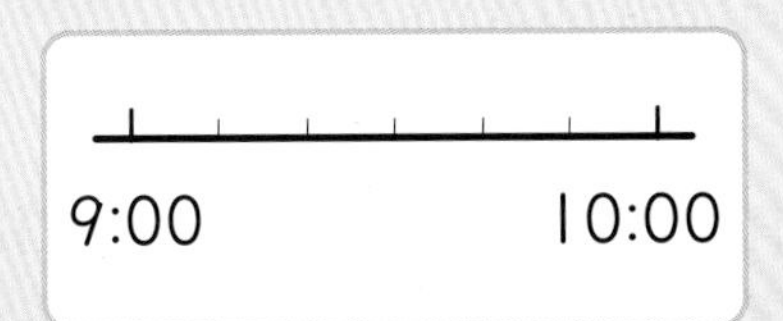

Haruto

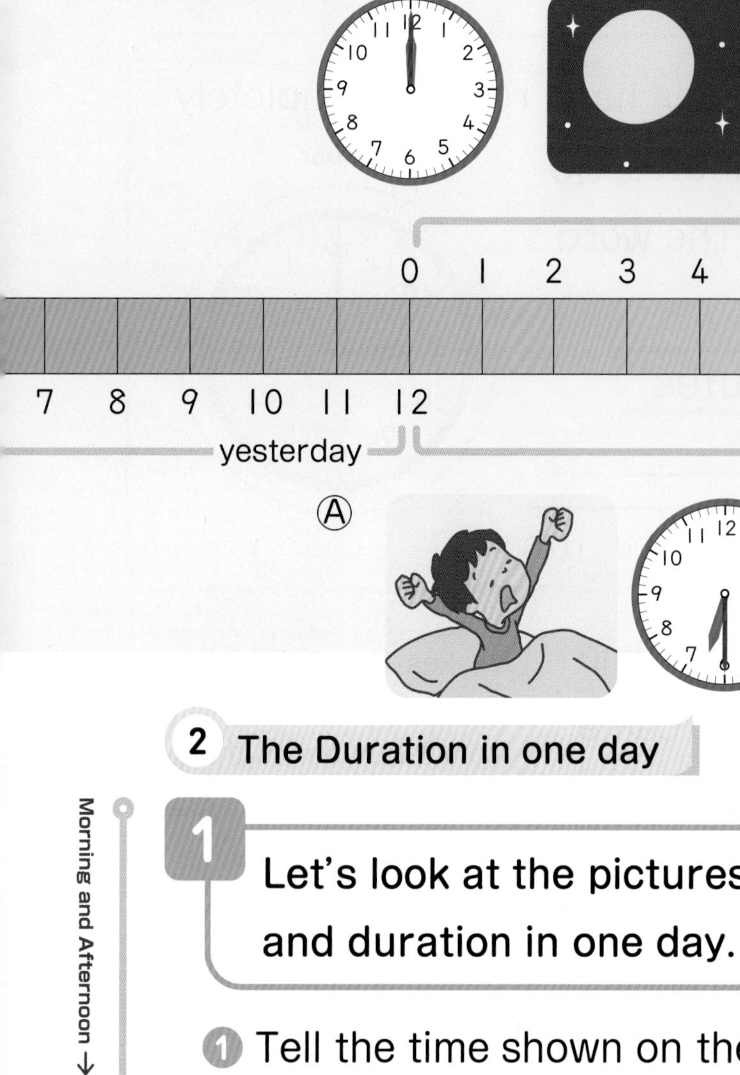

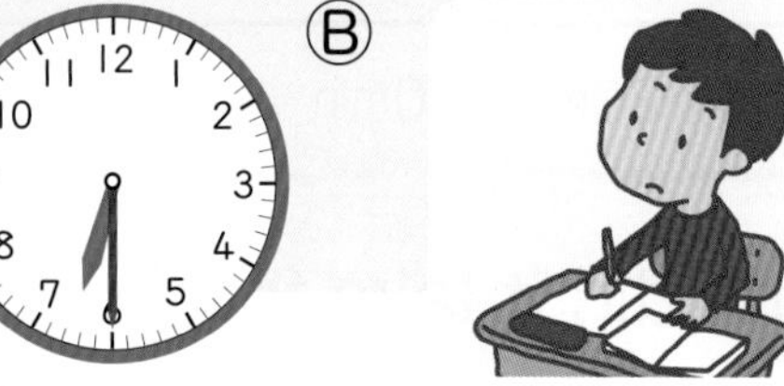

2 The Duration in one day

1 Let's look at the pictures above and think about time and duration in one day.

① Tell the time shown on the clocks Ⓐ, Ⓑ, Ⓒ, and Ⓓ above. Say whether it is "in the **morning**" or "in the **afternoon**".

\ Want to know /

? **Purpose** How many hours do we have in one day?

② How many hours do we have in the morning and in the afternoon?

→ morning … ☐ hours.

→ afternoon … ☐ hours.

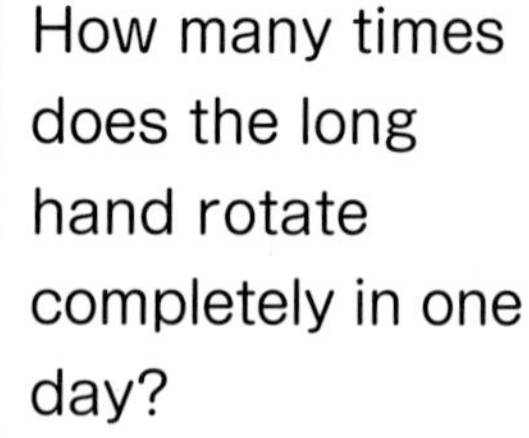

How many times does the long hand rotate completely in one day?

③ How many hours do we have in one day?

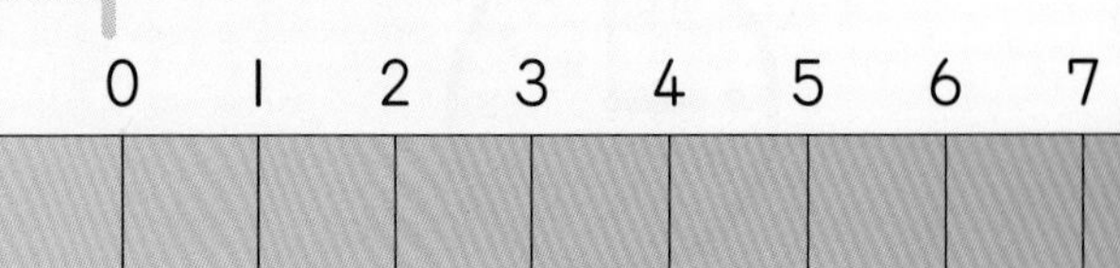

Ⓓ

Each day starts at 12 o'clock midnight. The short hand rotates completely around the clock twice a day.

"12 o'clock midnight" is the same as "12 o'clock in the afternoon". "12 o'clock noon" is the same as "12 o'clock in the morning."

Summary

In one day, morning has 12 hours. Afternoon has 12 hours. One day has 24 hours.

| 1 day = 24 hours |

12 o'clock midnight

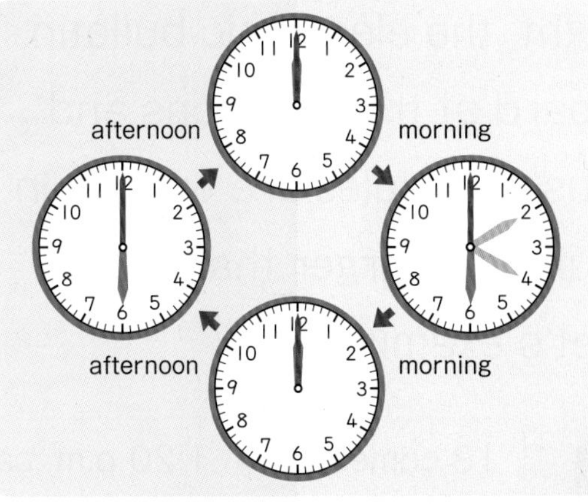

12 o'clock noon (noon)

That's it! **Morning / Afternoon**

"ante meridian" is a Latin word...

"a.m.," which stands for ante meridien, means before noon. 9:00 in the morning can be written as 9:00 a.m.
"p.m.", which stands for post meridian, means after noon (afternoon). 2:00 in the afternoon can be written as 2:00 p.m.

1 I went to bed at 9:00 p.m. and got up at 6:00 a.m. the next day. How long did I sleep?

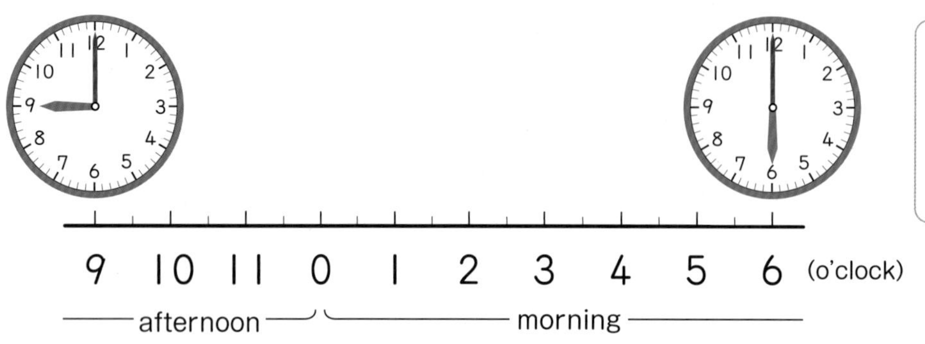

You slept 3 hours by 12 o'clock midnight.

2 Let's talk about how you spend your day using "morning" and "afternoon".

12:00 p.m. can be written as 0:00 a.m. next day.

That's it!

Schedules for Trains and Buses

As shown in the picture on the right, the electronic bulletin board at train stations and bus schedules are written in numbers larger than 12. Let's examine.

Electric bulletin board at a train station

Time	Mondays to Fridays	Saturdays and Sundays
11	07 22 32 47 57	14 40
12	07 22 32 47 57	14 40
13	07 22 32 47 57	14 40
14	07 22 32 47 57	14 40
15	07 22 32 47 57	14 40

Bus schedule

13 comes after 12...

1:20 p.m. can be written as 13:20.

Yu

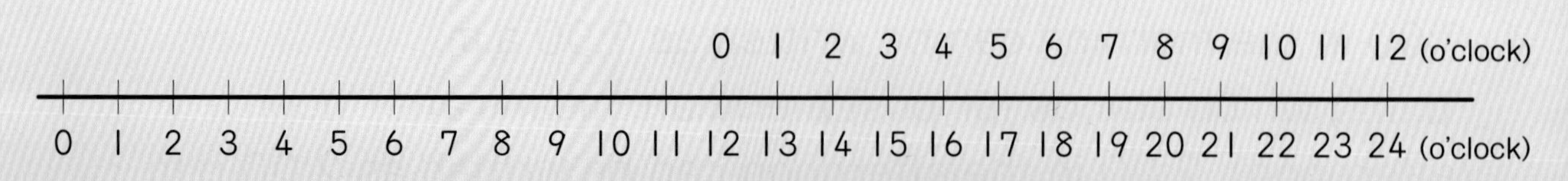

CAN What can you do?

☐ We understand the relationship between day, hour, and minute. → pp.23 ~ 25

1 Let's fill in each ☐ with numbers.

① 1 day= ☐ hours　② 60 minutes= ☐ hour

③ The duration for the long hand to rotate completely around the clock is ☐ hour.

④ The short hand rotates completely around the clock ☐ times a day.

☐ We understand the relationship between a.m. and p.m. → pp.24 ~ 25

2 Let's fill in each ☐ with words.

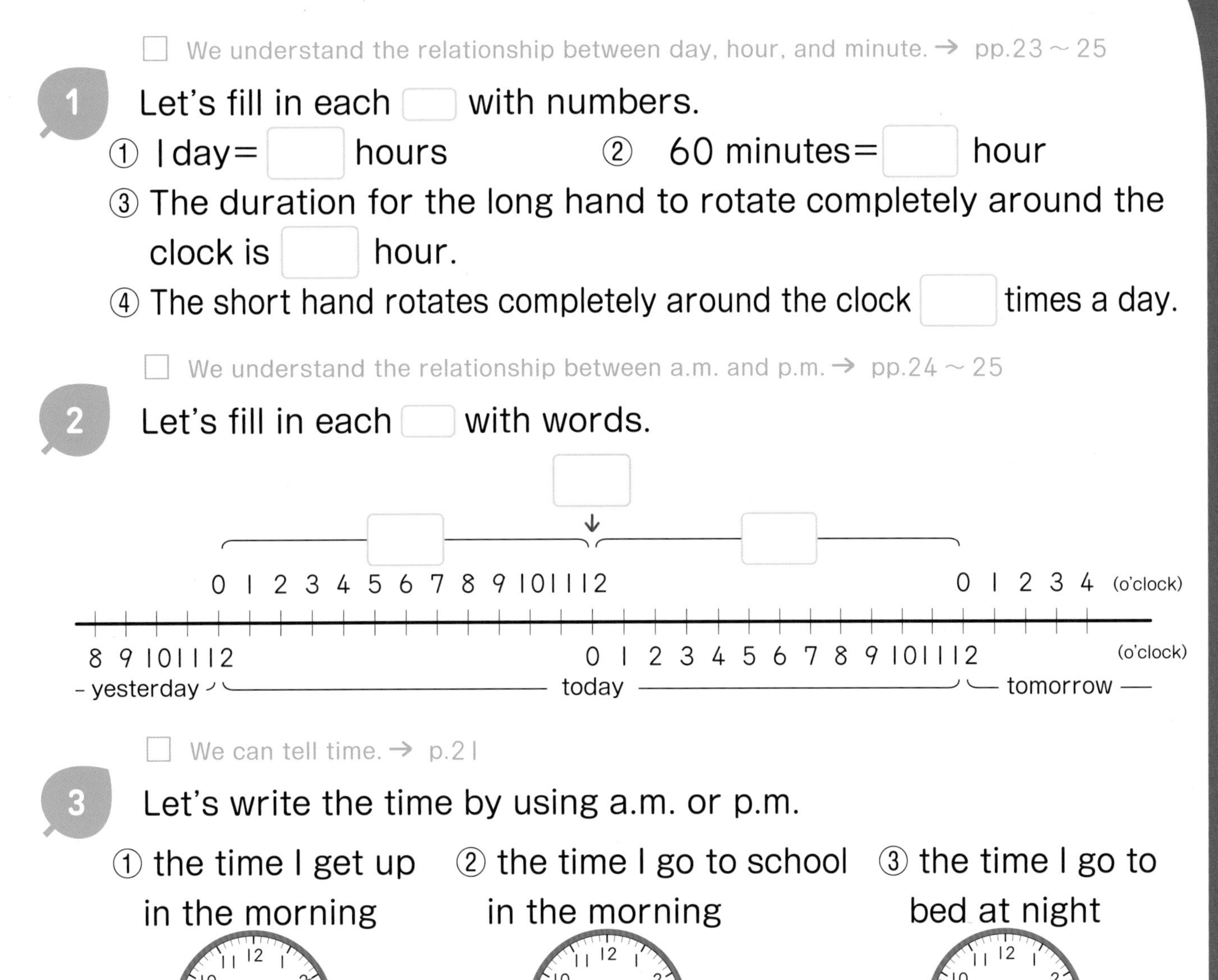

☐ We can tell time. → p.21

3 Let's write the time by using a.m. or p.m.

① the time I get up in the morning　② the time I go to school in the morning　③ the time I go to bed at night

Supplementary Problems → p.151

Which "Way to See and Think Monsters" did you find in 2 Time and Duration (1)" ?

I found "Unit" when I was thinking about the mark of the clock.

What other monsters did you find?

Usefulness and Efficiency of Learning

1 Let's draw the long hand of the clock that shows the time in the story.

I went to school at 7:30 a.m.

I ate lunch at 0:20 p.m.

I studied until 2:25 p.m.

2 Which of the following is the correct way to use words for "time" or "duration".

①

The duration we spent playing was 1 hour.

②

The duration of the beginning of the day is 0:00 a.m.

③

The duration of the meeting is 3:00 p.m.

④

The bus departed on time.

With the Way to see and think Monsters...

Let's Reflect!

Let's reflect on which monster you found in "2 Time and Duration (1)."

Unit

If we think of 60 minutes as one unit and express it as 1 hour, we can express one day as 24 hours.

① Let's reflect on time and duration.

1 mark of the clock represents ☐ minute.

Haruto

Sara

1 hour is 60 minutes, so the long hand moves ☐ in 1 hour.

Other way

If the clock is represented by a line of numbers, it is easy to understand the time.

② The clock is represented by a line of numbers below. Fill in the time for ↑.

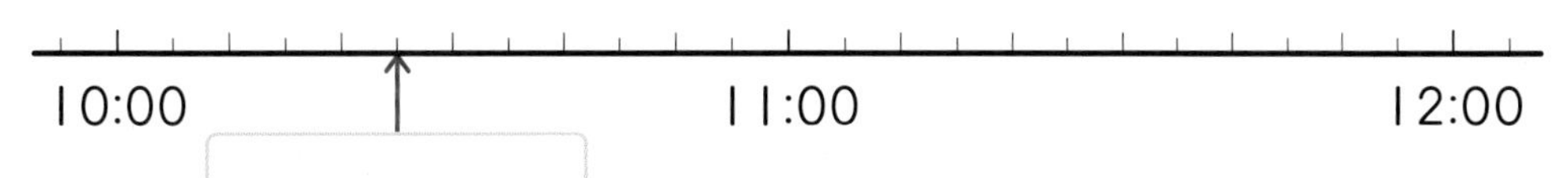

Solve the ?

By using the clock, I was able to find the relationship between the time and duration.

Akari

Want to Connect

If we use the clock, can we tell how many minutes ago or how many minutes after?

Yu

How many altogether? How many increase?

Let's think about how to find out the number of cookies.

Addition and Subtraction of 2-digit Numbers

Let's think about how to calculate easily.

1 Addition

Addition of 2-digit Numbers →

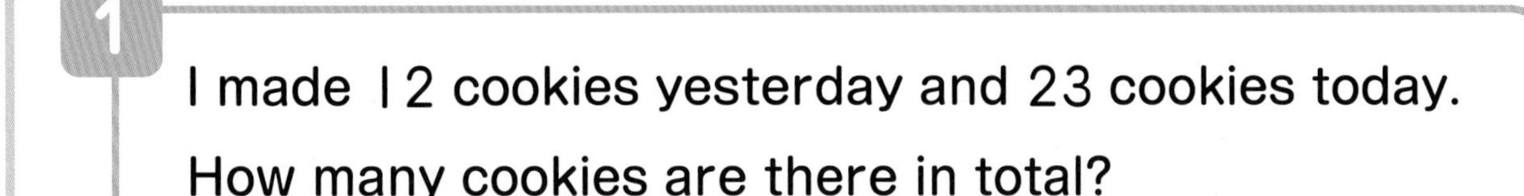

1
I made 12 cookies yesterday and 23 cookies today.
How many cookies are there in total?

① Let's write a math expression to find out the total number of cookies.

The number in each place is called a digit. 12 is a 2-digit number.

② Let's think about how many cookies there are in total. 

Can I find an easier way?

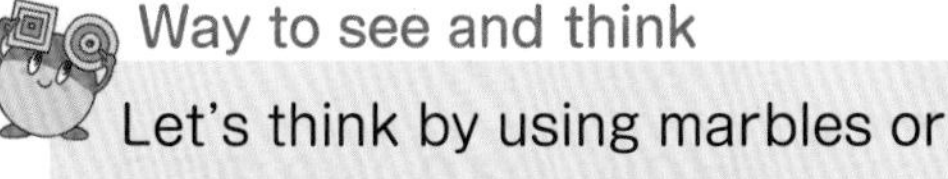

Let's think by using marbles or blocks to represent cookies.

③ What is similar among Akari, Haruto, and Sara's idea?

Akari's idea

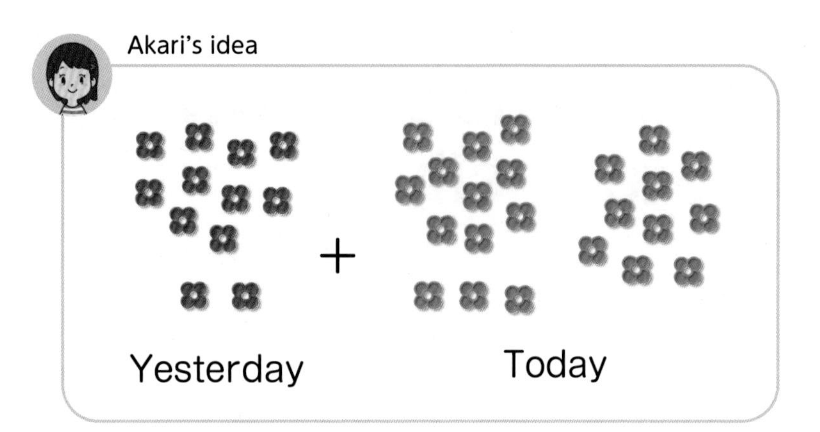

If I replace the cookies with marbles...

Haruto's idea

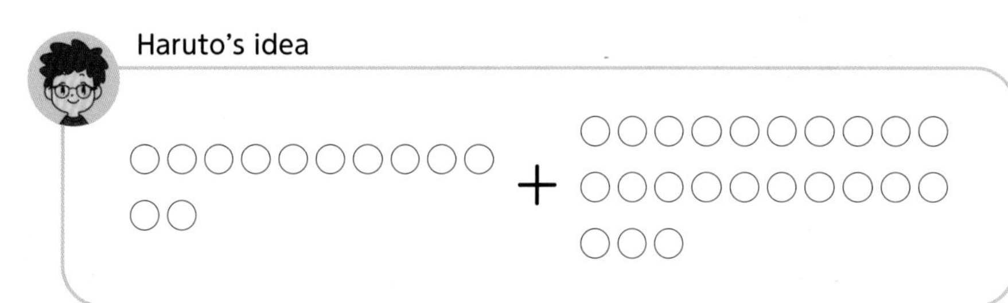

I replaced the cookies using ○ and lined them up.

Sara's idea

If we think by using the blocks...

Way to see and think

Sets of how many are used?

④ Let's think about how to calculate 12 + 23.

Remember we made sets of 10 to operate 1-digit numbers.

How should I operate 2-digit numbers?

Way to see and think

You operated the numbers in the ones place for 12 + 3.

Want to think

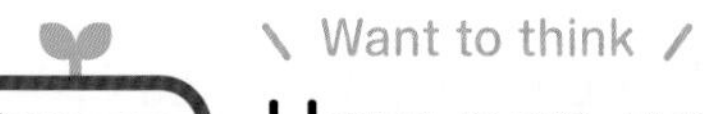

How can we add 2-digit numbers?

Yu's idea

12 + 23

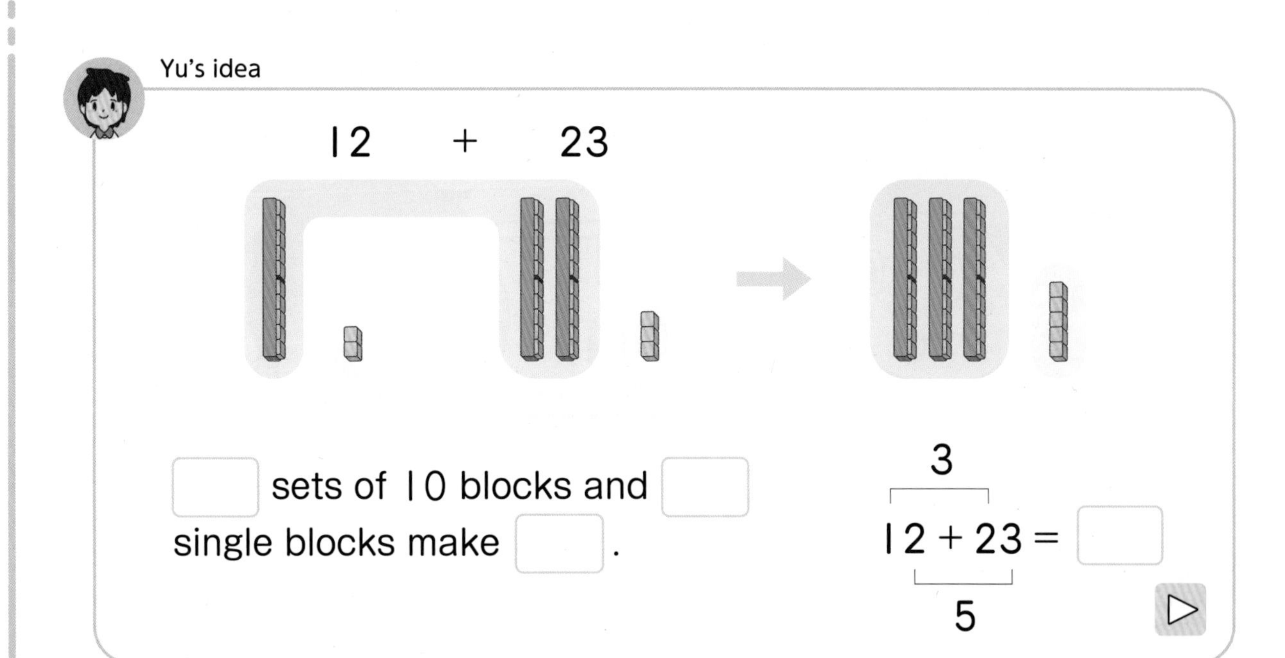

☐ sets of 10 blocks and ☐ single blocks make ☐.

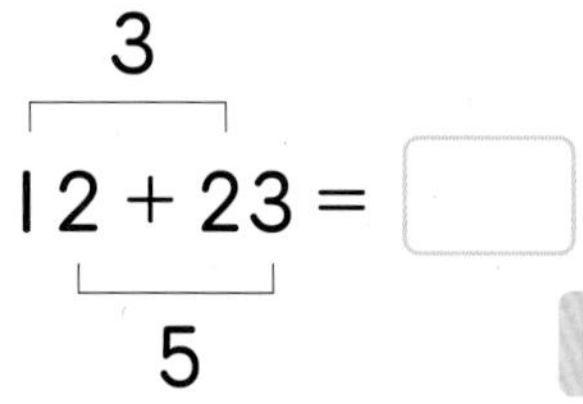

Sara's idea

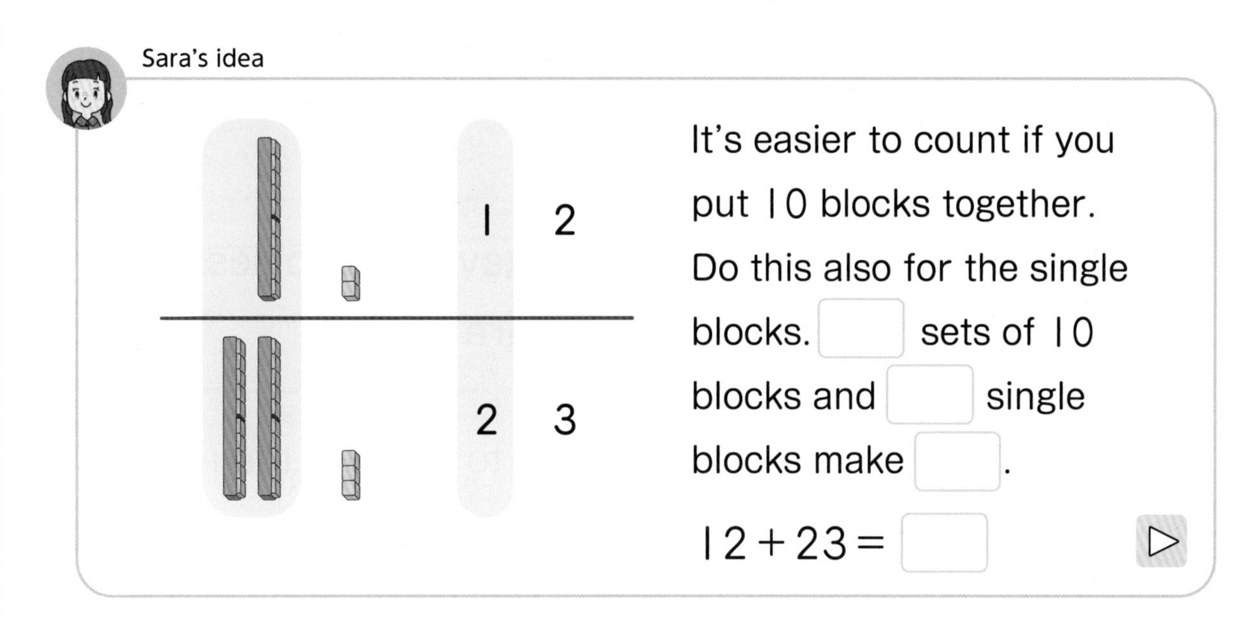

It's easier to count if you put 10 blocks together. Do this also for the single blocks. ☐ sets of 10 blocks and ☐ single blocks make ☐.

$12 + 23 = \square$

Math Sentence: $12 + 23 = 35$

Answer: 35 cookies

Summary

For adding 2-digit numbers, add the numbers in the ones place and then add the numbers in the tens place.

? Can we also do subtractions of 2-digit numbers in the same way?

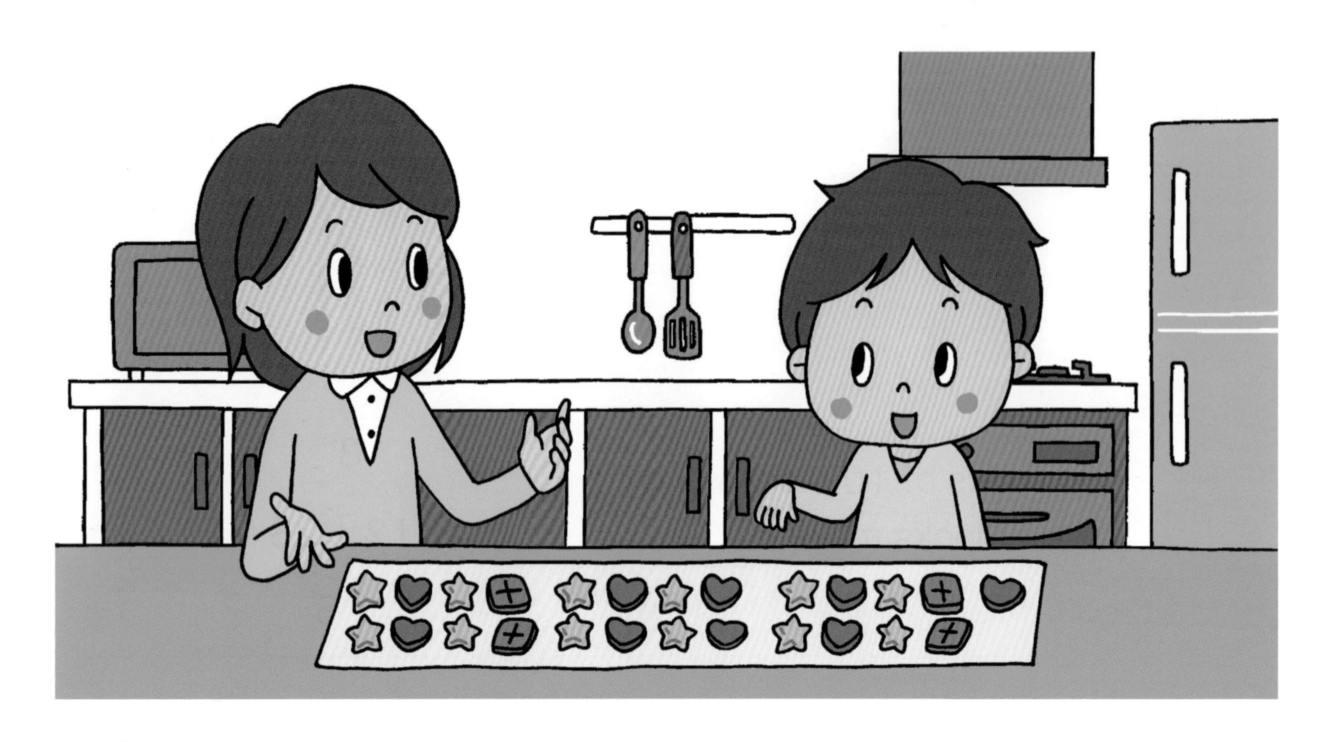

2 Subtraction

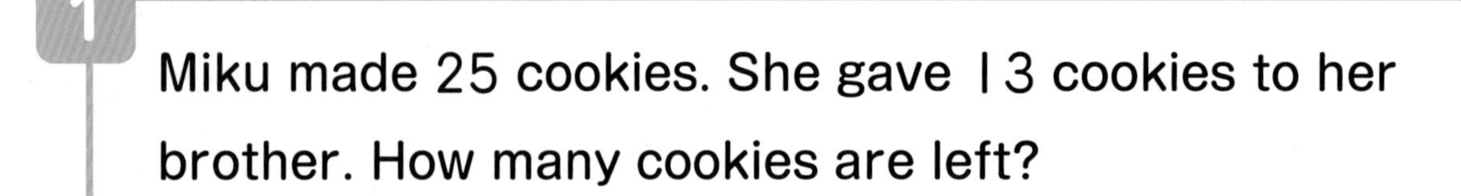

1 Miku made 25 cookies. She gave 13 cookies to her brother. How many cookies are left?

① Let's write a math expression to find out the number of the cookies left.

② Let's think about how many cookies are left.

Yu

Can we make it easier like we did in addition?

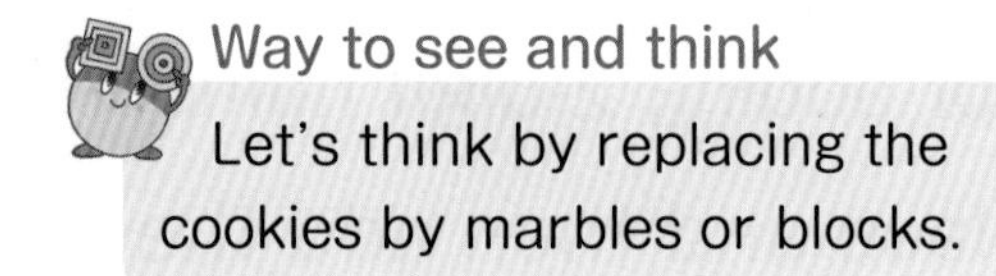

Way to see and think
Let's think by replacing the cookies by marbles or blocks.

③ What is similar among Akari, Haruto, and Sara's idea?

Akari's idea

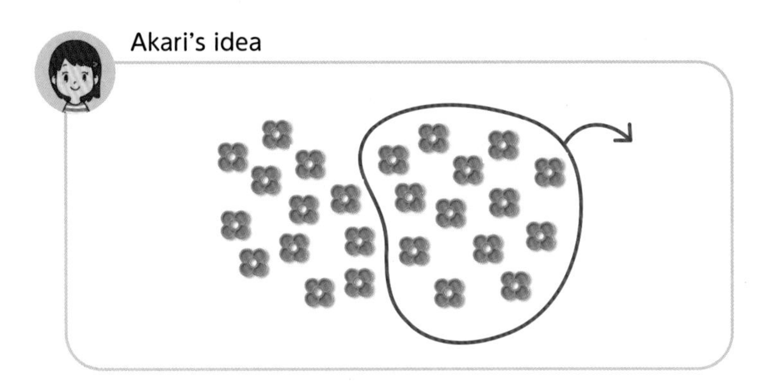

If we replace the cookies with marbles...

Akari

Haruto's idea

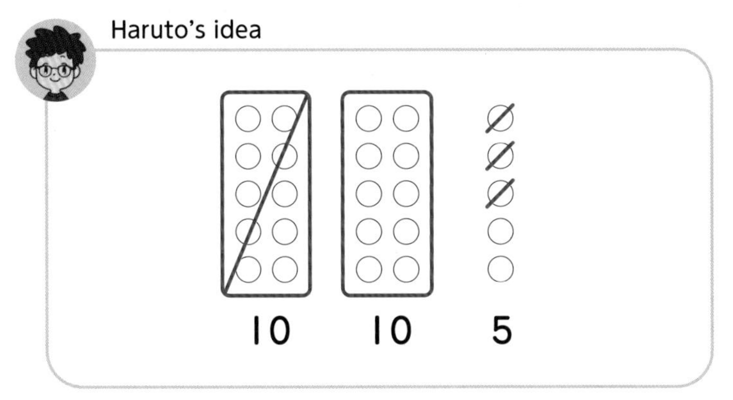

If we replace the cookies with ○ and make sets of 10...

Haruto

Sara's idea

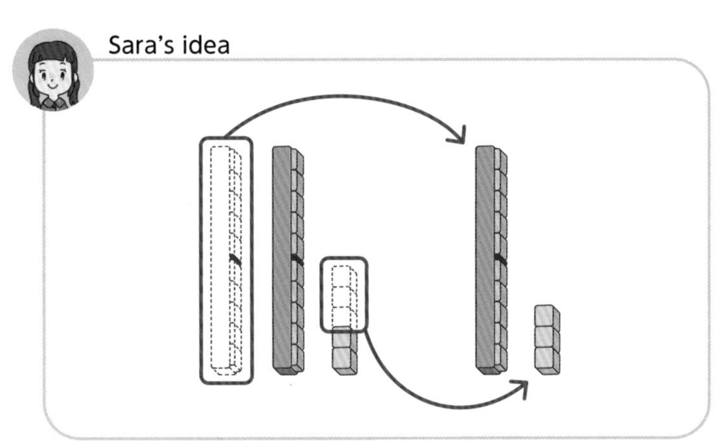

If we think by using blocks...

Sara

Way to see and think

Sets of 10 are used in all ideas...

4 Let's think about how to calculate for finding the answer for 25 − 13.

Yu

Remember we decomposed a 2-digit number into 10 and some in the 1st grade.

Can we subtract 2-digit numbers?

Akari

Way to see and think

You calculated the numbers in the ones place for 25 − 3.

Want to think

? How can we subtract 2-digit numbers?

Yu's idea

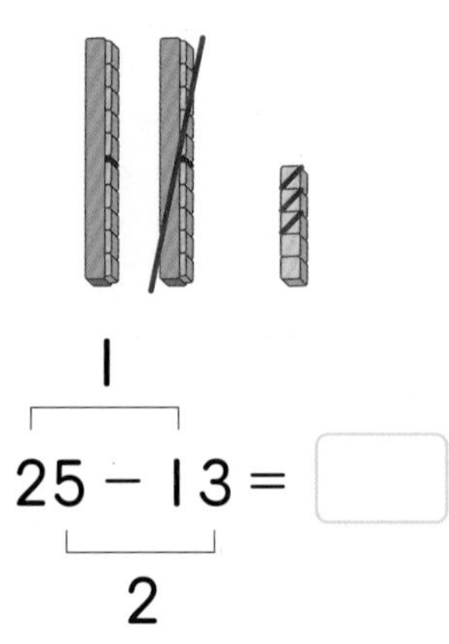

Decompose 25 into [ ] and 5.

Decompose 13 into 10 and [ ].

$20 - 10 =$ [ ].

$5 - 3 =$ [ ].

[ ] and 2 make [ ].

Sara's idea

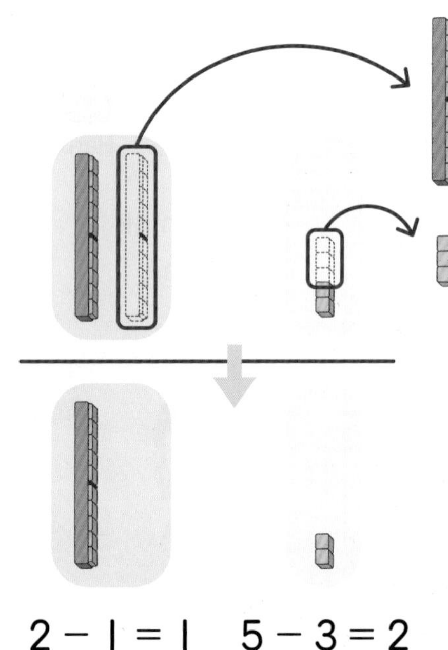

There are [ ] sets of 10.

[ ] taken away 1 is [ ].

5 taken away [ ] is [ ].

[ ] is in the tens place and

[ ] is in the ones place.

So, the difference is [ ].

Math Sentence: $25 - 13 = 12$

Answer: 12 cookies

For subtracting 2-digit numbers, subtract the numbers in the ones place and then subtract the numbers in the tens place.

With the Way to see and think Monsters...

Let's Reflect!

Let's reflect on which monster you found in "3 Addition and Subtraction of 2-digit Numbers."

Divide

Same Way

By dividing the tens place into sets and singles, we could calculate addition and subtraction in the same way as we learned in 1st Grade.

① In what way did you think to find out the answer for the calculation of 2-digit numbers?

[12 + 23]

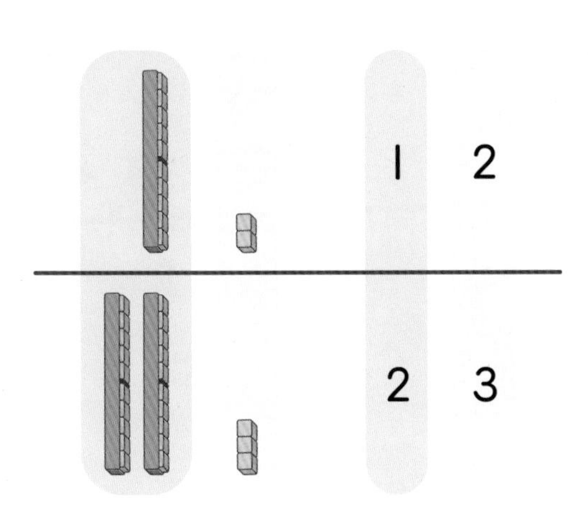

[25 − 13]

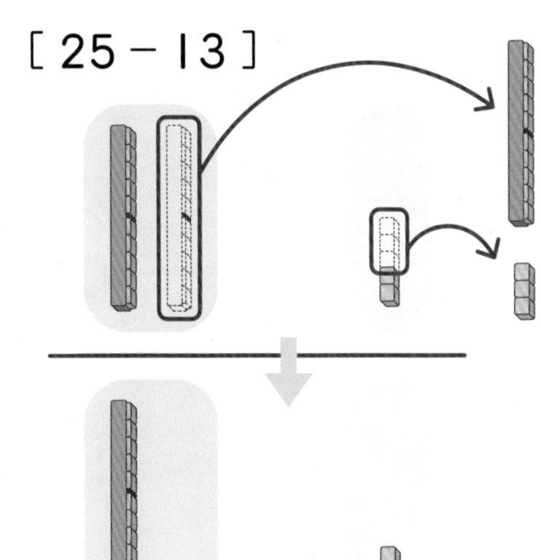

I replaced the 2-digit numbers into blocks, and thought by making sets of 10 blocks and single blocks.

Haruto

In both addition and subtraction, it was easier to think by separating the 2-digits into tens place and ones place.

Akari

Solve the ?

We could find out the answer to the 2-digit addition and subtraction by using blocks and by separating into tens place and ones place.

Yu

Want to Connect

Can't we calculate without using blocks?

Sara

Find the ?

How many bloomed in total?

So many tulips bloomed!
There are 24 red tulips.
There are 13 yellow ones.
How many tulips are there in total?

Let's think about how to add two 2-digit numbers.

Addition in Vertical Form

Let's think about the meaning of addition and how to add.

1 Addition of 2-digit numbers

1

Tulips have bloomed. There are 24 red tulips and 13 yellow tulips.
How many tulips are there in total?

in total

24 red tulips | 13 yellow tulips

① Let's write a math expression.

② Let's think about how to add.

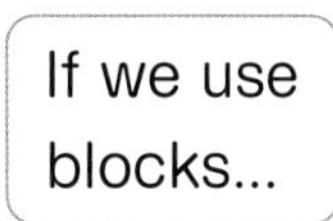

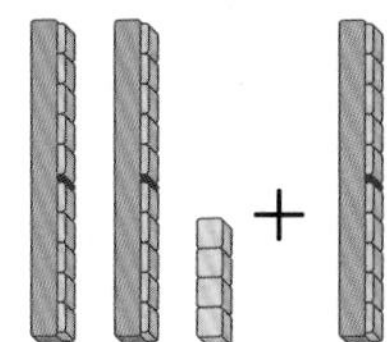

24 + 13 can be written vertically by aligning the digits of the numbers according to their places. Such algorithm is called addition in **vertical form**.

$$\begin{array}{r} 24 \\ +\ 13 \\ \hline \end{array}$$

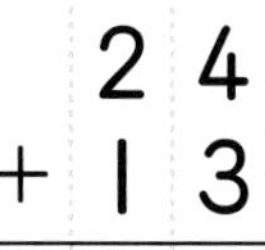

Want to know

Purpose How do we add in vertical form?

Addition algorithm for 24 + 13 in vertical form

Tens	Ones
(2 tens)	(4 ones)
(1 ten)	(3 ones)

↓

(3 tens)	(7 ones)

Tens	Ones
2	4
+ 1	3

Align the digits of the numbers according to their places.

We also separated the tens and ones when we used blocks!
Akari

↓

	Tens	Ones
	2	4
+	1	3
	3	7

$2+1=3$ $4+3=7$

Add the numbers in the ones place and then add the numbers in the tens place.

Math Sentence: $24+13=37$

Answer: 37 tulips

1 Let's calculate the following in vertical form.

①

	3	1
+	5	7

②

	1	7
+	3	0

③ $15+62$

④ $65+31$

⑤ $18+40$

⑥ $32+20$

⑦ $50+36$

⑧ $20+70$

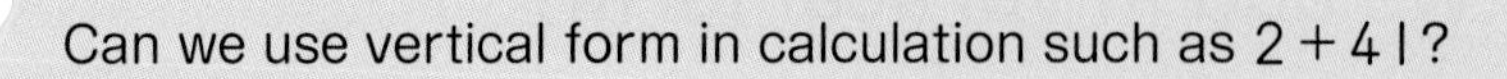

?
Can we use vertical form in calculation such as 2 + 41?

2 Let's think about how to add 2 + 41 in vertical form.

① Which is the correct way of writing the numbers?

Let's think about the approximate answer to 2 + 41.

Yu's idea

$$\begin{array}{rcc} & 2 & \\ + & 4 & 1 \\ \hline \end{array}$$

② Let's add in vertical form.

+		

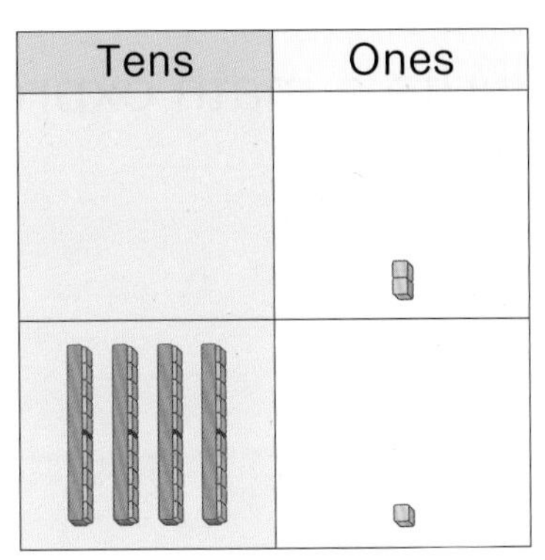

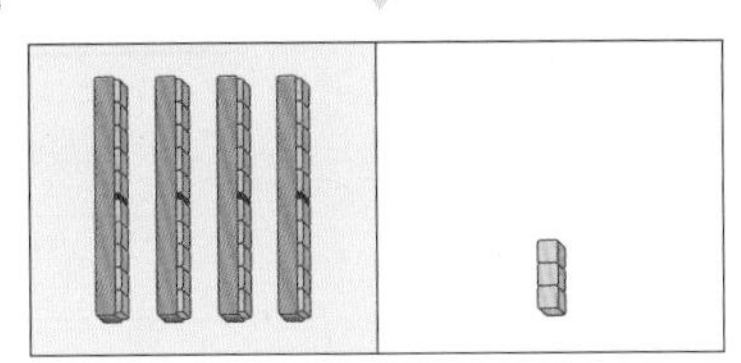

Way to see and think

It's easier to think by replacing the numbers with blocks.

If we align the ones and tens, the way of calculation is the same as what we have learned so far.

Haruto

Summary Way to see and think

For adding in vertical form, align the digits of the numbers according to their places and then add the numbers in the same places.

1 Let's calculate the following in vertical form.

① 4 + 23　② 7 + 82　③ 91 + 8　④ 65 + 3

? What should I do when the answer to the addition of the ones place is more than 10?

3

There are 38 picture books and 27 reference books in Sakura's classroom.
How many books are there in total?

in total

38 picture books　27 reference books

① Let's write a math expression.

② Let's think about how to add in vertical form.

	3	8
+	2	7

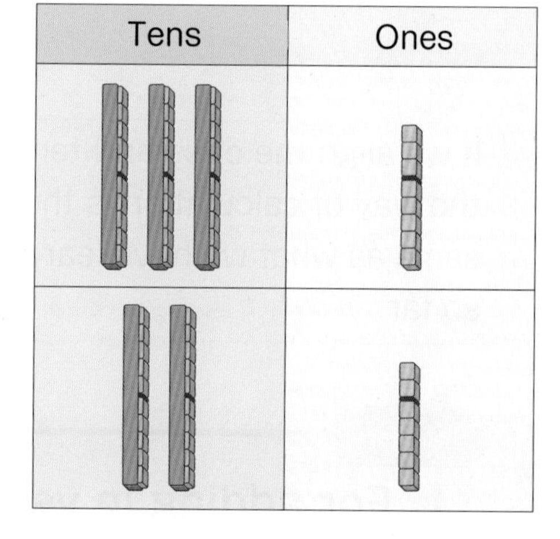

Haruto: What is the difference from the calculation of 24 + 13?

Akari: In the ones place, 8 + 7 = 15. So...

\ Want to know /

When the answer to the addition of the ones place is more than 10, how can we add in vertical form?

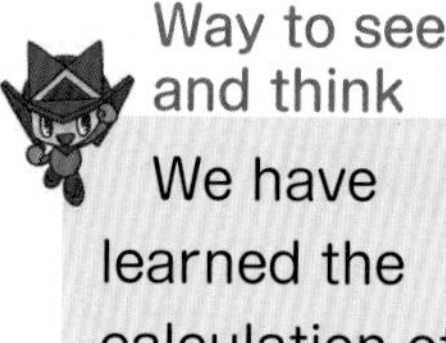

Way to see and think

We have learned the calculation of 8 + 7 in the 1st grade.

③ Let's explain Haruto and Akari's ideas.

Haruto's idea

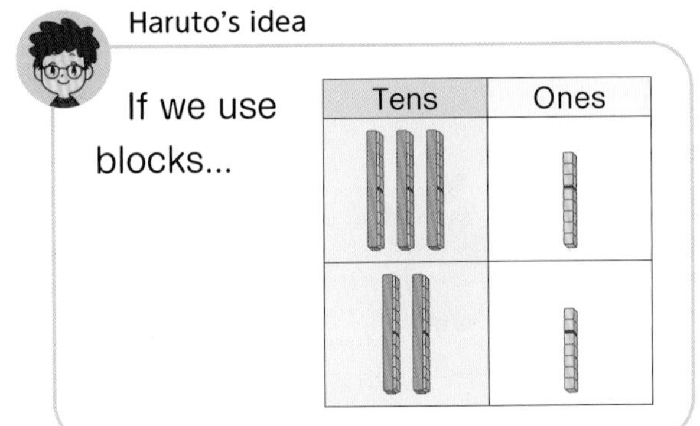

Akari's idea

$$\begin{array}{r} 38 \\ +\,27 \\ \hline 15 \\ +\,50 \\ \hline 65 \end{array}$$

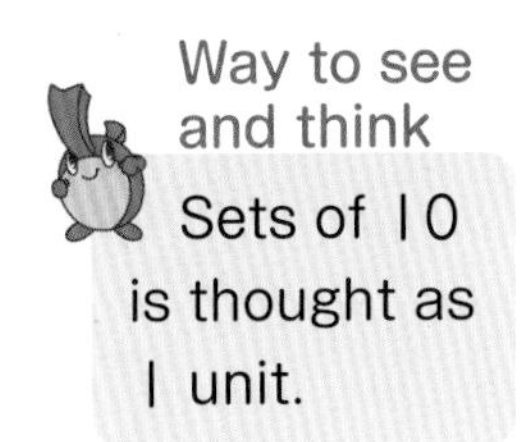

Way to see and think
Sets of 10 is thought as 1 unit.

If the sum is more than 10 when we add the numbers in a place, the set of 10 is moved to the next higher place. This is called **carrying** or **regrouping**.

Addition algorithm for 38 + 27 in vertical form

Tens	Ones

$$\begin{array}{r} 38 \\ +\,27 \\ \hline \end{array}$$

Align the digits of the numbers according to their places.
Add the numbers in the ones place first.

Ones Place

$8 + 7 = 15$

The ones place is ☐.
Carry 1 ten to the tens place.

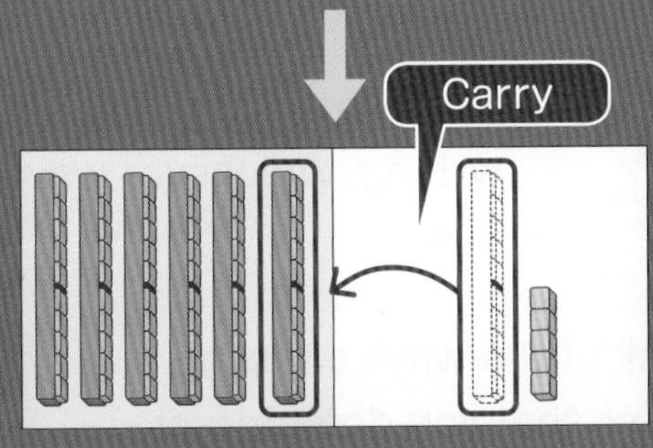

$$\begin{array}{r} 38 \\ +\,27 \\ \hline 1 \\ 65 \end{array}$$

Tens Place

1 ten was carried, so

$3 + 2 + 1 = 6$

The tens place is ☐.

Math Sentence: $38 + 27 = 65$ Answer: 65 books

1 Let's calculate the following in vertical form paying attention to the carrying,

① 27 + 65 ② 48 + 34

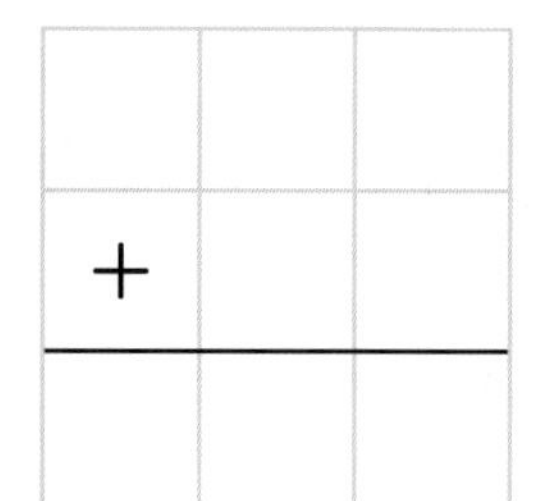

+

In the case of addition with carrying, let's start calculating from the ones place.

Summary

If the sum is more than 10 when we add the numbers in the ones place, carry 10 ones to the tens place as 1 ten.

2 Let's calculate the following in vertical form.

① 28 + 16 ② 47 + 27 ③ 59 + 37 ④ 15 + 56

⑤ 43 + 38 ⑥ 18 + 78 ⑦ 24 + 19 ⑧ 49 + 13

? Can we do more addition using the vertical form?

That's it! How to write the carried number in vertical form

$$\begin{array}{r} {\scriptstyle 1}\ \ \\ 38 \\ +\,27 \\ \hline 65 \end{array} \qquad \begin{array}{r} 3\ \ 8 \\ +\,2\ \ 7 \\ \hline 6\,{\scriptstyle 1}\,5 \end{array}$$

There are many ways of writing. Whichever way is fine.

Think of the ways of doing this so that you don't forget the carried 1.

4 Find the mistakes in the following processes. Let's talk about how to calculate correctly.

① $27+53$

$$\begin{array}{r} 27 \\ +\ 53 \\ \hline 70 \end{array}$$

② $35+6$

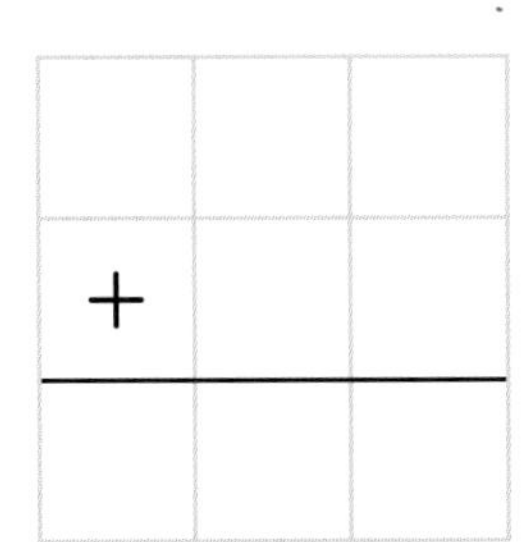

③ $7+23$

$$\begin{array}{r} 7 \\ +\ 23 \\ \hline 20 \end{array}$$

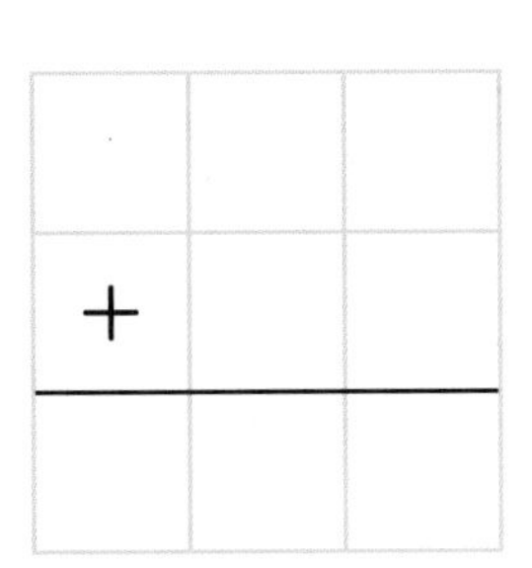

Be careful when carrying.
Yu

\ Want to discuss /

Purpose What should we pay attention to when we calculate in vertical form?

1 Let's calculate the following in vertical form.

① $72+18$ ② $35+45$ ③ $16+24$ ④ $33+17$

⑤ $54+7$ ⑥ $77+9$ ⑦ $6+89$ ⑧ $5+15$

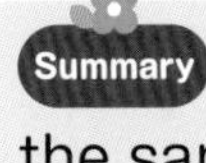

Summary When you calculate, be careful to align the same digits, and not to forget carrying.

2 The answer of the addition sentence below is 50. Let's make a math sentence.

$\square + \square = 50$

We can make many math sentences.
Akari

? The answer for $20+30$ and $30+20$ are both 50, but would it be the same in other numbers even when the order of the augend and the addend is changed?

2 Rules of addition

There are 38 strawberries in a box and 16 strawberries in a basket.
How many strawberries are there in total?

in total

38 in a box | 16 in a basket

① Let's make a math expression based on Haruto and Sara's idea.

Haruto's idea
We put the strawberries in the basket into the box.
augend
addend
+

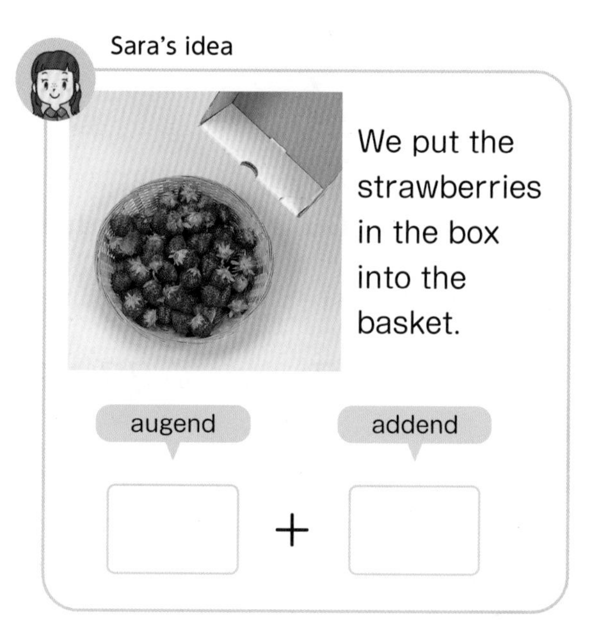

Sara's idea
We put the strawberries in the box into the basket.
augend
addend
+

The order of the two numbers has changed.

Both of them are expressing the same situation, so...

\ Want to know /

Are the answers the same even when the order of the augend and the addend are changed?

② Let's check if the answers to 38+16 and 16+38 is the same by using the vertical form.

	3	8
+	1	6

	1	6
+	3	8

Can we do the same with the addition of 1-digit?
Haruto

1 Let's check if the answers for addition using other numbers will be the same even when the order of the augend and the addend is changed.

In addition, the answers are the same even when the order of the augend and the addend is changed.

augend		addend		augend		addend
38	+	16	=	16	+	38

38	16
16	38

Way to see and think

2 Let's find the math expressions with the same answers.

24 + 31, 70 + 11, 37 + 21, 21 + 37, 62 + 9, 9 + 62, 72 + 9, 31 + 24

Can you find some without calculation?

? Are there other rules of addition?

2

I had 32 marbles. I received 7 marbles from my brother and 3 marbles from my sister.
How many marbles do I have in total?

❶ Let's write a math expression.

\ Want to think /
Purpose How can we calculate easily?

Haruto

❷ Let's think about how to calculate.

Yu's idea

Since I received 7 marbles from my brother,
$32 + 7 = 39$
Since I received 3 marbles from my sister,
$39 + 3 = 42$

Akari's idea

I combined the marbles I received from my brother and sister,
$7 + 3 = 10$
Because I already had 32 marbles,
$32 + 10 = 42$

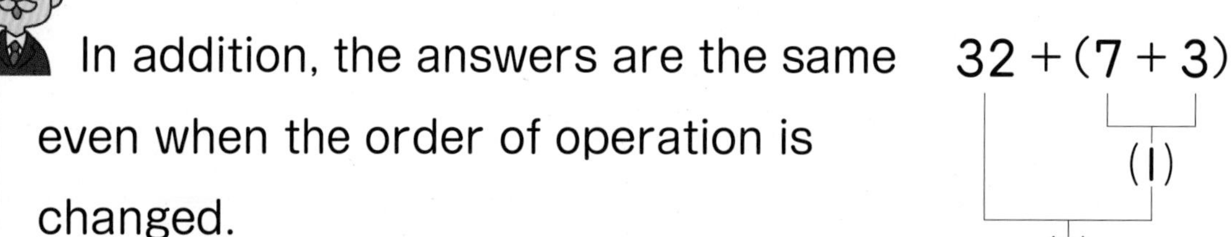

In addition, the answers are the same even when the order of operation is changed.

$32 + (7 + 3)$
(1) $7 + 3$
(2) $32 + (7 + 3)$

$$(32 + 7) + 3 = 32 + (7 + 3)$$

(　　) is the symbol that indicates we calculate first.

1 Let's find an easier way to add.

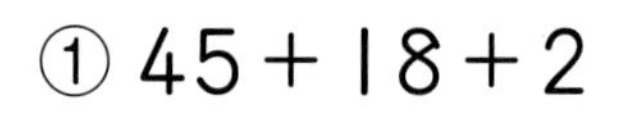

① $45 + 18 + 2$　② $58 + 13 + 27$

③ $68 + 13 + 12$　④ $44 + 37 + 6$

Which combination of numbers makes addition simpler?

CAN What can you do?

☐ We understand the meaning of the way to add in vertical form. → p.43

1 Let's summarize how to add $67+28$ in vertical form.

① In the ones place, $7+8$ makes 15.

The ones place is ☐.

Carry ☐ ten to the tens place.

② In the tens place, $6+2+$ ☐ $=9$

③ The answer is ☐.

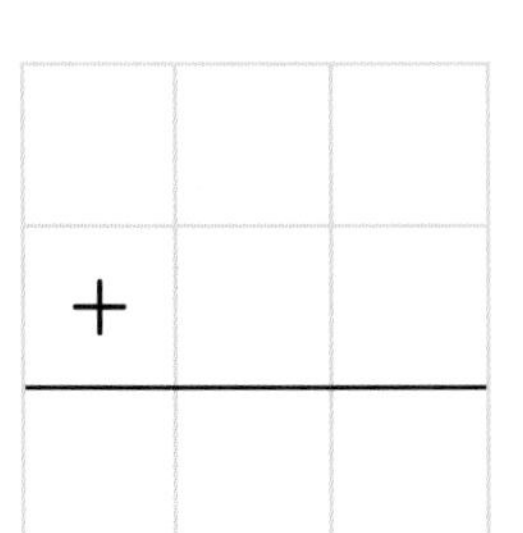

☐ We can add in vertical form. → pp.39 ~ 45

2 Let's calculate the following in vertical form.

① $36+32$ ② $43+34$ ③ $2+55$ ④ $40+47$

⑤ $38+25$ ⑥ $57+19$ ⑦ $35+58$ ⑧ $17+43$

⑨ $18+9$ ⑩ $49+4$ ⑪ $8+47$ ⑫ $5+75$

☐ We can find the answers by making addition expressions. → pp.39 ~ 45

3 In Itsuki's school, there are 2 classes for the 2nd grade. There are 31 children in the 1st class and 29 children in the 2nd class. How many children are there in total in the 2nd grade?

Supplementary Problems → p.152

Which "Way to See and Think Monsters" did you find in 4 Addition in Vertical Form"?

Yu: I found "Align" when I was calculating addition in vertical form.

Akari: I found other monsters too!

Utilize Usefulness and Efficiency of Learning

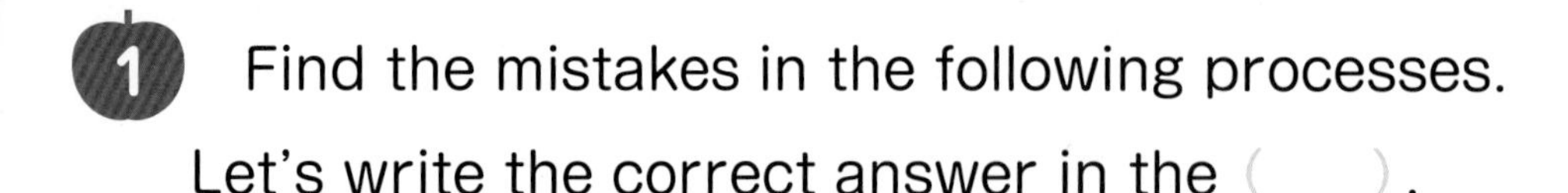

1 Find the mistakes in the following processes.

Let's write the correct answer in the (   ).

① $\begin{array}{r} 27 \\ +\ 43 \\ \hline 60 \end{array}$ (   )

② $\begin{array}{r} 6 \\ +\ 35 \\ \hline 95 \end{array}$ (   )

③ $\begin{array}{r} 18 \\ +\ 39 \\ \hline 47 \end{array}$ (   )

2 Hiroshi folded cranes on Friday, Saturday, and Sunday.

Let's answer the following questions.

Friday	Saturday	Sunday
I folded ☐ cranes.	I folded 8 more cranes than yesterday.	I folded 38 more cranes than yesterday.

① If Hiroshi folded 24 cranes on Friday, how many cranes did he fold on Saturday and Sunday?

② Based on the pictures above, Chiharu made the following math expression.

$8 + 38$

What is Chiharu trying to find out? Explain in your words.

With the Way to see and think Monsters...

Let's Reflect!

Let's reflect on which monster you found in "4 Addition in Vertical Form."

Align

For addition in vertical form, we could calculate by aligning the digits of the numbers according to their places and then adding the numbers in the same place.

① Is the addition in vertical form written on the right correct? Also, what do we need to be careful when we add in vertical form?

$$\begin{array}{r} 2 \\ +\ 4\ 7 \\ \hline 6\ 7 \end{array}$$

We need to align the digits of the numbers according to their places, so we need to write 2 in the ☐ place.

Sara

Rule

We found out the rules of addition by trying addition with various numbers.

② What rule did you find out in addition?

○ In addition, the answers are the ☐ even when the order of the augend and the addend are changed.

augend		addend		augend		addend
38	+	16	=	16	+	38

○ In addition, the answers are the ☐ even when the order of operation is changed. $(32+7)+3=32+(7+3)$

Solve the ?

For adding 2-digit numbers in vertical form, we can calculate by adding the numbers in the ones place and then the numbers in the tens place.

Haruto

Want to Connect

Can we calculate in the same way even when the numbers get larger?

Akari

Find the

How many are left?

Let's think about how many pieces are left.

5 Subtraction in Vertical Form

Let's think about the meaning of subtraction and how to subtract.

1 Subtraction of 2-digit numbers

1

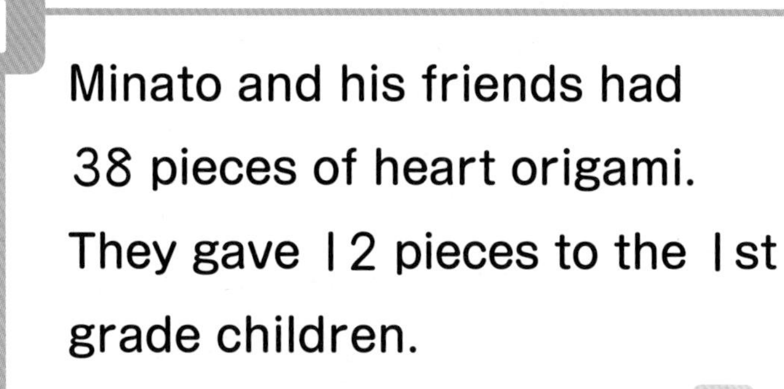

Minato and his friends had 38 pieces of heart origami. They gave 12 pieces to the 1st grade children.

How many pieces are left? ▷

38 pieces

12 pieces given　　the number of pieces left

① Let's write a math expression.

② Let's think about how to subtract 38 − 12 in vertical form.

	3	8
−	1	2

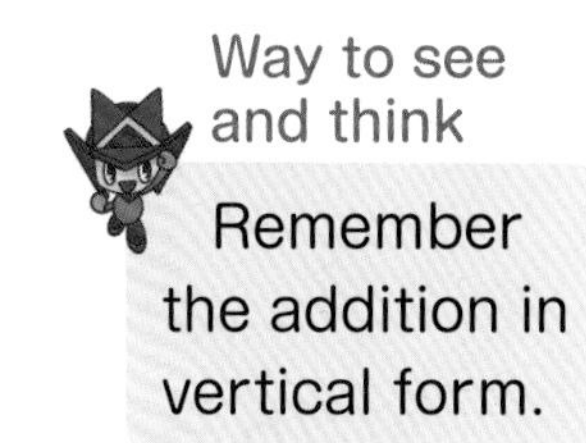

Way to see and think
Remember the addition in vertical form.

? Purpose ＼ Want to know ／ How do we subtract in vertical form?

Subtraction algorithm for 38 − 12 in vertical form

Tens	Ones

Tens | Ones

$$\begin{array}{r} 38 \\ -\ 12 \\ \hline \end{array}$$

Align the digits of the numbers according to their places.

Way to see and think

Same as addition in vertical form, we calculate the ones place and the tens place separately.

$$\begin{array}{r} 38 \\ -\ 12 \\ \hline 26 \end{array}$$

Subtract the numbers in the ones place and then subtract the numbers in the tens place.

3 − 1 = 2 | 8 − 2 = 6

Math Sentence: 38 − 12 = 26 Answer: 26 pieces

1 Let's calculate the following in vertical form.

①

	4	7
−	2	5

②

	8	8
−	5	1

We can calculate either from the ones place or the tens place.

③ 76 − 32 ④ 59 − 45 ⑤ 36 − 24

⑥ 49 − 13 ⑦ 97 − 76 ⑧ 66 − 25

? Can we use vertical form in any subtractions?

2 Let's think about how to calculate the following in vertical form.

❶ 34 − 14　　❷ 68 − 64　　❸ 54 − 40

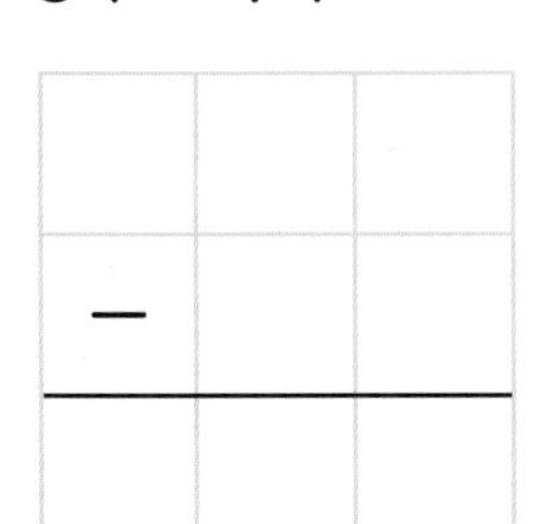

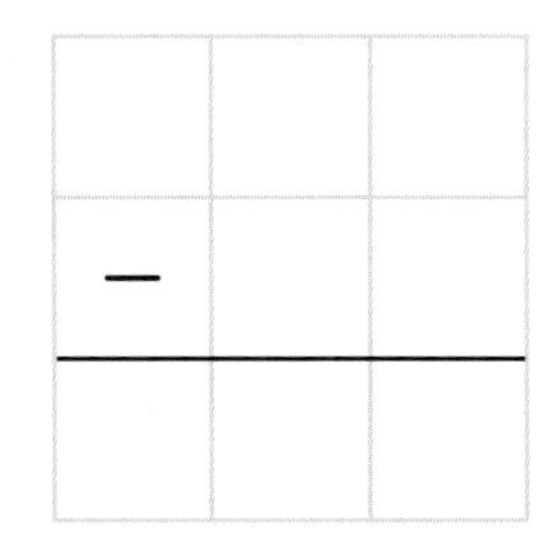

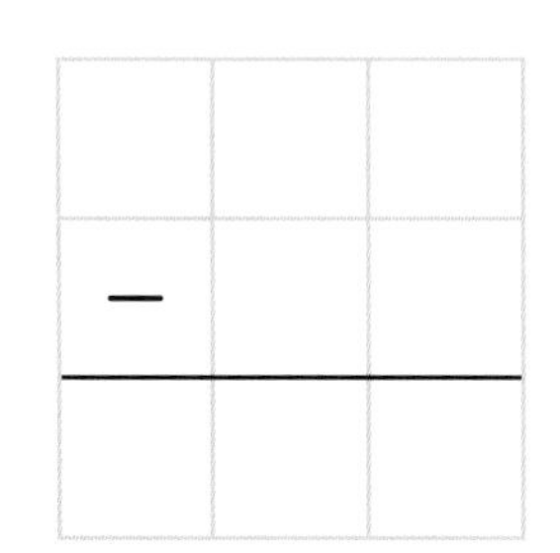

1 Let's think about how to calculate the following in vertical form.

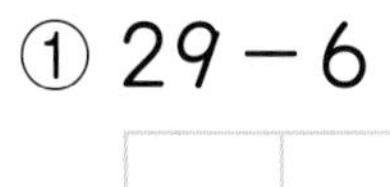

① 29 − 6

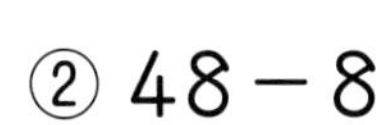

② 48 − 8

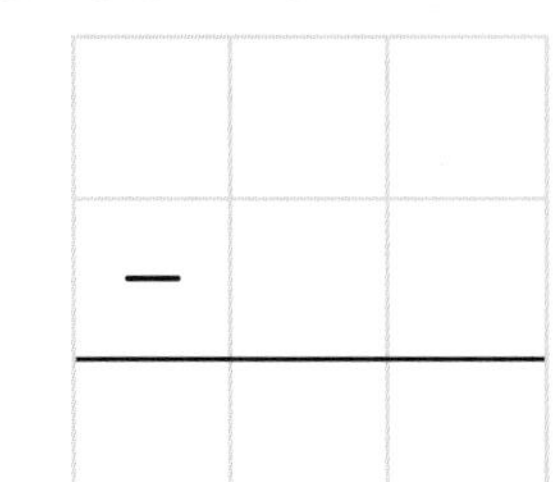

Way to see and think

When calculating in vertical form, we aligned the digits of the numbers according to their places.

Let's think about the approximate answer to 29 − 6.

Summary

For subtracting in vertical form, align the digits of the numbers according to their places and then subtract the numbers in the same place.

2 Let's calculate the following in vertical form.

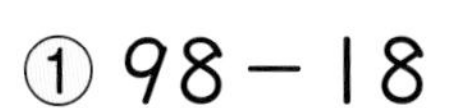

① 98 − 18　　② 43 − 42　　③ 30 − 20

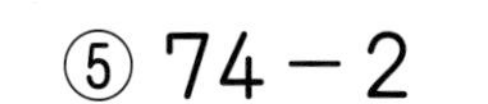

④ 58 − 5　　⑤ 74 − 2　　⑥ 85 − 5

? What should I do when I can't subtract in the ones place?

3

There were 45 sheets of origami paper. I used 27 sheets. How many sheets are left?

Even if we don't have ○ in the diagram, we can think about the approximate answer.

❶ Let's write a math expression.

❷ Let's think about how to calculate in vertical form.

	4	5
−	2	7

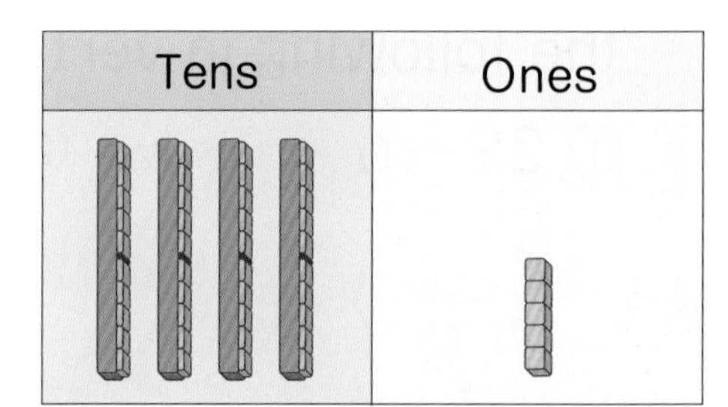

What is the difference between the calculation of 38 − 12?

Yu

In the ones place, we can't do 5 − 7...

Sara

\ Want to know /

? Purpose **When the subtraction of ones place cannot be done, how can we subtract in vertical form?**

Way to see and think

We have learned the calculation of 15 − 7 in the 1st grade.

When the subtraction of the ones place cannot be done, borrow 1 ten from the tens place as 10 ones. This is called **borrowing** or **regrouping**.

Subtraction algorithm for 45 − 27 in vertical form 

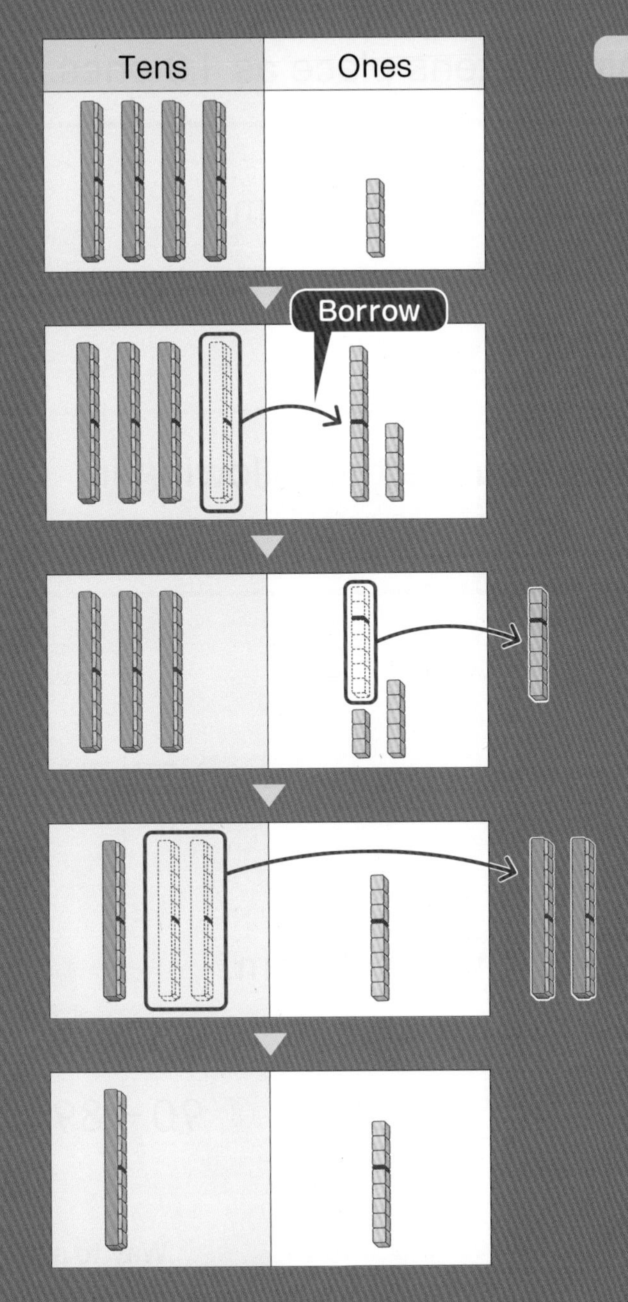

Tens Ones

$$\begin{array}{r} 45 \\ -\ 27 \\ \hline \end{array}$$

Align the digits of the numbers according to their places.

Ones Place

$$\begin{array}{r} \overset{3}{\cancel{4}}\overset{10}{5} \\ -\ 27 \\ \hline 8 \end{array}$$

Borrow 1 ten from the tens place as 10 ones, so 15 − 7 = 8.
The ones place of the answer becomes ☐.

Tens Place

$$\begin{array}{r} \overset{3}{\cancel{4}}\overset{10}{5} \\ -\ 27 \\ \hline 18 \end{array}$$

1 ten has been moved to the ones place, so 3 − 2 = ☐.
The tens place of the answer becomes ☐.

$$\begin{array}{r} 45 \\ -\ 27 \\ \hline 18 \end{array}$$

Math Sentence: 45 − 27 = 18 Answer: 18 sheets

1 Let's calculate 53 − 36 paying attention to the borrowing.

Summary

When the subtraction of the ones place cannot be done, borrow 1 ten from the tens place as 10 ones.

2 Let's calculate the following in vertical form.

① 41 − 19 ② 72 − 33 ③ 81 − 16

4

Let's think about how to calculate the following in vertical form.

❶ 70 − 43

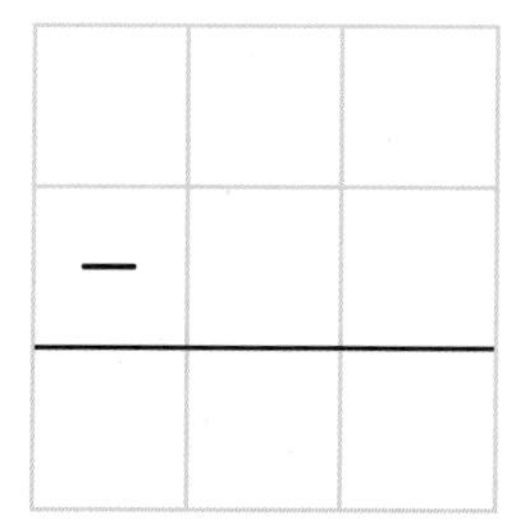

❷ 34 − 26

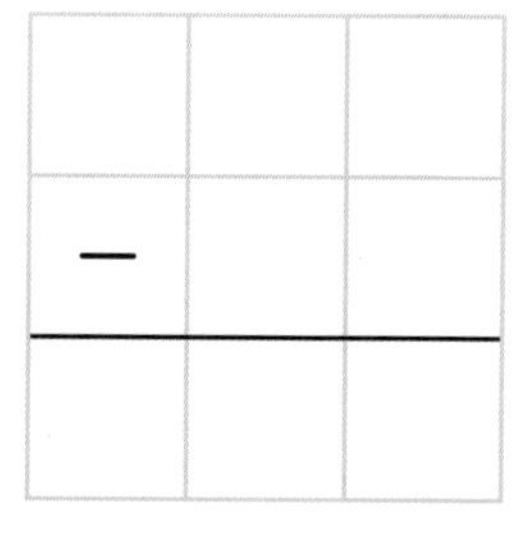

1 Let's calculate the following in vertical form.

① 70 − 56 ② 40 − 24 ③ 50 − 33 ④ 80 − 48

⑤ 26 − 18 ⑥ 54 − 45 ⑦ 73 − 67 ⑧ 90 − 89

2 Let's think about how to calculate 35 − 8 in vertical form.

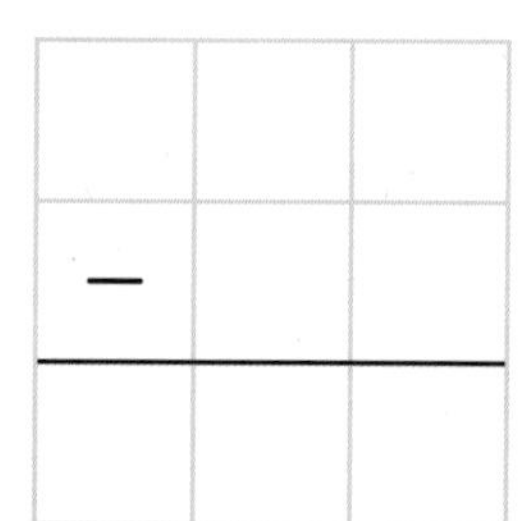

Way to see and think

Let's align the digits when you write.

3 Let's calculate the following in vertical form.

① 92 − 8 ② 51 − 9 ③ 40 − 7 ④ 60 − 3

Various subtraction in vertical form ↓

2 Relationship between addition and subtraction

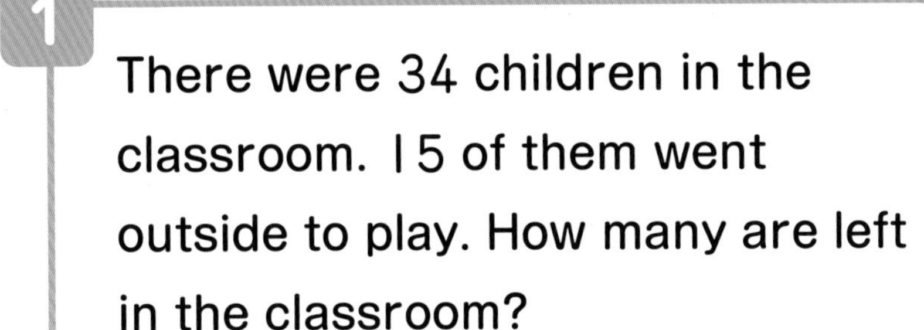

1 There were 34 children in the classroom. 15 of them went outside to play. How many are left in the classroom?

34 children in the classroom at first

15 children went outside　　☐ children were left

① Let's write a math expression, and find out the answer.

☐　　Answer: ☐ children:

Did I get the answer correctly?

How can I check whether I am doing it correctly?

Want to solve

? **Purpose** Can we confirm whether the answer to the subtraction is correct?

② When the 15 children who went outside come back to the classroom, how many will be there?

I will add 15 children who came back to those who are already in the classroom, so...

Yu

③ Let's compare the math sentence you made in ① and ②, and talk about them with your friends.

1 Let's confirm whether the calculation of $41-28=13$ is correct or not by calculating $13+28$.

In subtraction, the number that you add the subtrahend and the answer will be the same as the minuend. This method can be used to confirm the answer in subtraction.

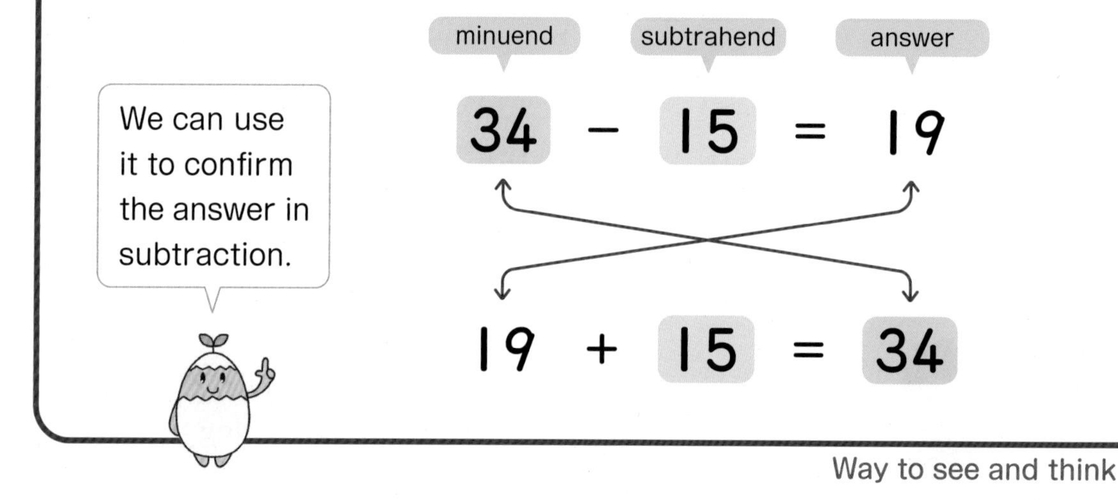

2 Which of the math expressions Ⓐ, Ⓑ, Ⓒ, or Ⓓ would be the one to confirm the answers for the following subtractions? Choose one for each of them.

① $76-51$
② $32-26$
③ $45-8$
④ $50-7$

Ⓐ $37+8$
Ⓑ $6+26$
Ⓒ $25+51$
Ⓓ $43+7$

I think I can find out the answer by calculating either addition or subtraction.

CAN What can you do?

☐ We understand the meaning of the way to subtract in vertical form. → p.57

1 Let's summarize how to subtract 73 − 26 in vertical form.

① Borrow ☐ ten from the tens place, so that the ones place will become ☐ − 6 = ☐.

② In the tens place, ☐ − 2 = ☐.

③ The answer is ☐.

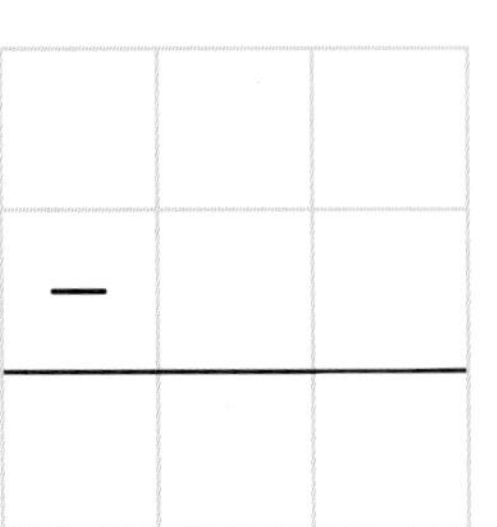

☐ We can subtract in vertical form and confirm the answer. → pp.53 ~ 60

2 Let's calculate the following in vertical form, and confirm the answer.

① 58 − 32 ② 66 − 23 ③ 33 − 11 ④ 28 − 12

⑤ 87 − 19 ⑥ 63 − 24 ⑦ 42 − 13 ⑧ 80 − 17

⑨ 50 − 49 ⑩ 34 − 27 ⑪ 44 − 38 ⑫ 64 − 8

☐ We can find out the answers by making subtraction expressions. → pp.53 ~ 60

3 Koharu had 32 candies. She gave 14 of them to her brother. How many candies are left?

Supplementary Problems →p.153

Which "Way to See and Think Monsters" did you find in 5 Subtraction in Vertical Form"?

Yu: I found "Align" when I was calculating subtraction in vertical form.

Akari: I found other monsters too!

Utilize Usefulness and Efficiency of Learning

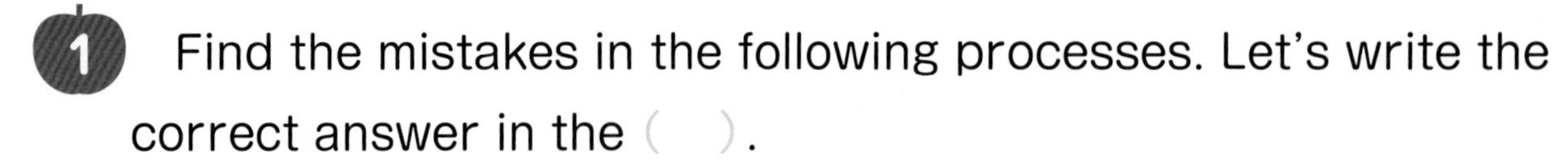

1 Find the mistakes in the following processes. Let's write the correct answer in the (　).

① $\begin{array}{r} 72 \\ -\ 47 \\ \hline 35 \end{array}$　　② $\begin{array}{r} 65 \\ -\ 43 \\ \hline 12 \end{array}$　　③ $\begin{array}{r} 58 \\ -\ 3 \\ \hline 28 \end{array}$

(　　　)　　(　　　)　　(　　　)

2 What numbers were eaten by the worms? Let's fill in each □ with a number.

(Example) $\begin{array}{r} \boxed{B}\ 5 \\ -\ 1\ \boxed{A} \\ \hline 7\ 1 \end{array}$

〈How to find out the answer〉

- A is the number that makes 5 − $\boxed{A}$ = 1, so A = 4.
- B is the number that makes $\boxed{B}$ − 1 = 7, so B = 8.

① 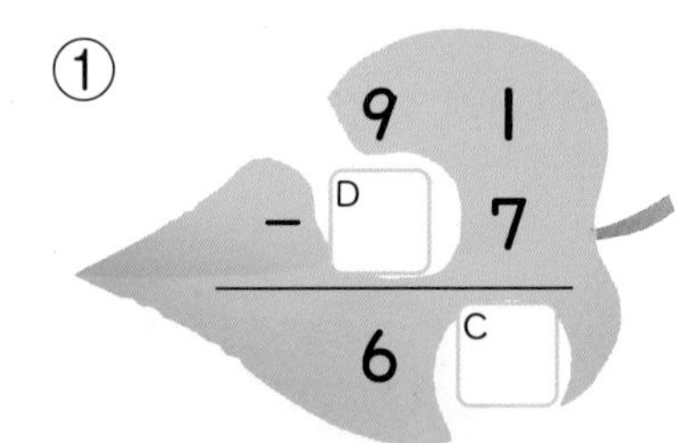

② 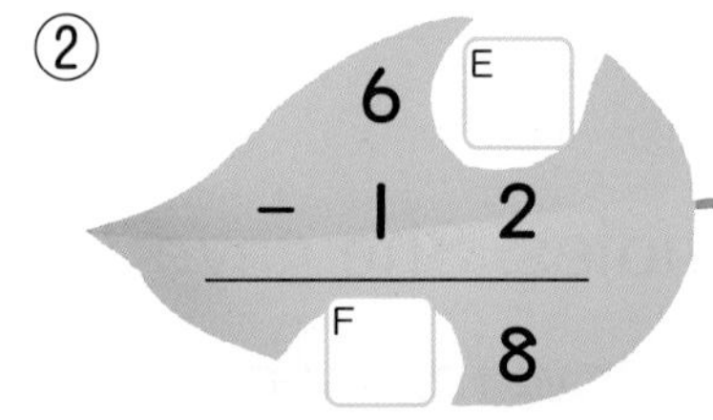

3 Make more problems like the ones above. Exchange them with your friends and solve them.

① Solve the problem correctly.

② Decide what numbers to replace with □.

③ Do the problem yourself to confirm if it can be solved.

(Example 1)

$\begin{array}{r} 3\ 8 \\ +\ 2\ 6 \\ \hline 6\ 4 \end{array}$ ▶ $\begin{array}{r} \square\ 8 \\ +\ 2\ \square \\ \hline 6\ 4 \end{array}$

(Example 2)

$\begin{array}{r} 8\ 7 \\ -\ 2\ 9 \\ \hline 5\ 8 \end{array}$ ▶ $\begin{array}{r} 8\ \square \\ -\ \square\ 9 \\ \hline 5\ 8 \end{array}$

With the Way to see and think Monsters...

Let's Reflect!

Let's reflect on which monster you found in "5 Subtraction in Vertical Form."

Align

For subtraction in vertical form, we could calculate by aligning the digits of the numbers according to their places and then subtracting the numbers in the same place.

① Is the subtraction in vertical form written on the right correct? Also, what do we need to be careful when we subtract in vertical form?

```
  6 3
- 5
-----
  1 3
```

We need to align the digits of the numbers according to their places, so we need to write 5 in the ☐ place.

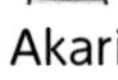

Akari

Rule

We found out the rules of subtraction by trying subtraction with various numbers.

② Let's fill in each ☐ on the right with a number.

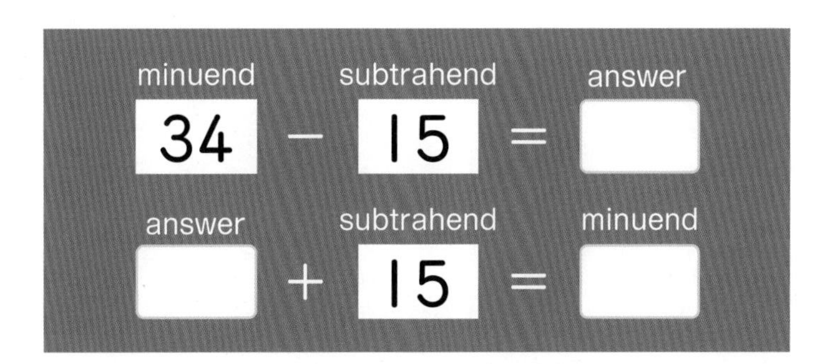

minuend		subtrahend		answer
34	−	15	=	☐
answer		**subtrahend**		**minuend**
☐	+	15	=	☐

Solve the ?

Same as the addition with 2-digit numbers, we can calculate subtraction with 2-digit numbers.

Sara

Want to Connect

Can we calculate in the same way even when the numbers get larger?

Haruto

Find the ?

Which is longer?

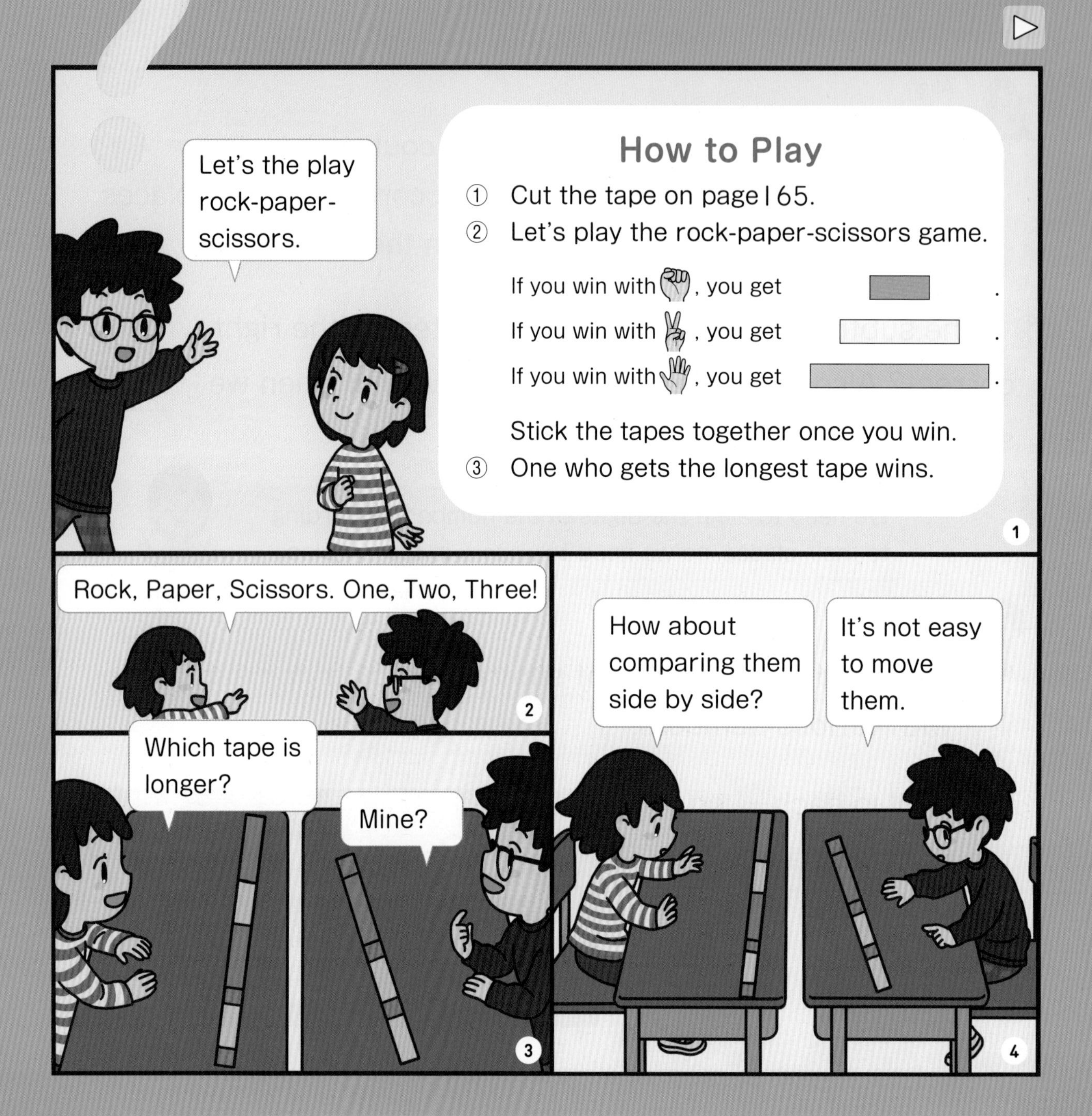

\ Want to compare /

Purpose Let's think about how to compare lengths.

6 Length (1)

Let's think about how to compare lengths and how to represent lengths.

1 How to compare lengths

How to compare lengths →

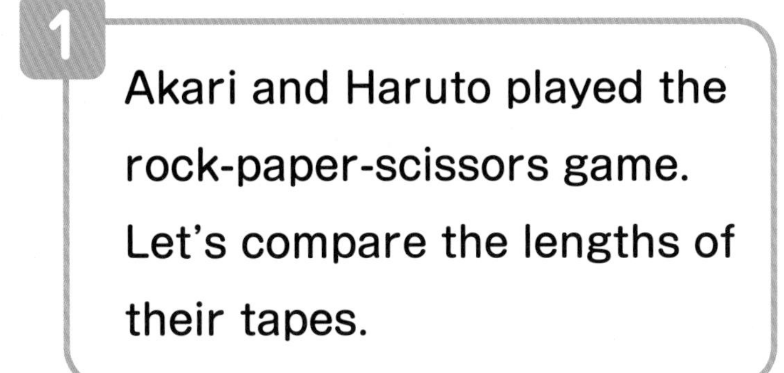

1
Akari and Haruto played the rock-paper-scissors game. Let's compare the lengths of their tapes.

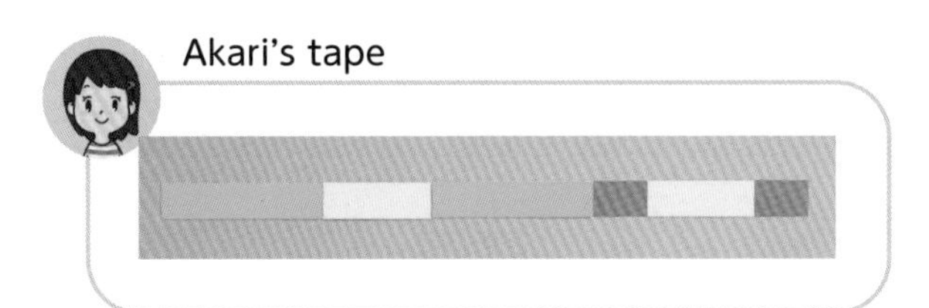

Akari's tape

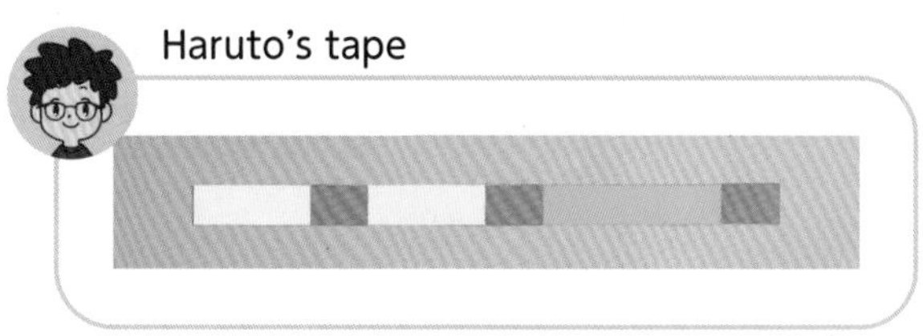

Haruto's tape

① Let's talk about how to compare the lengths.

How about aligning their edges?

We compared by counting the number of some units in the 1st grade.

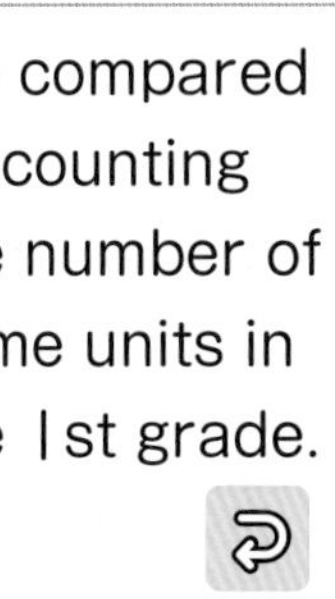

But we have to move them.

Then we can compare the lengths without moving them.

Way to see and think

Can you compare them in the same way as you have learned in the 1st grade?

❷ Let's represent the lengths by examining the number of the unit of the same tape.

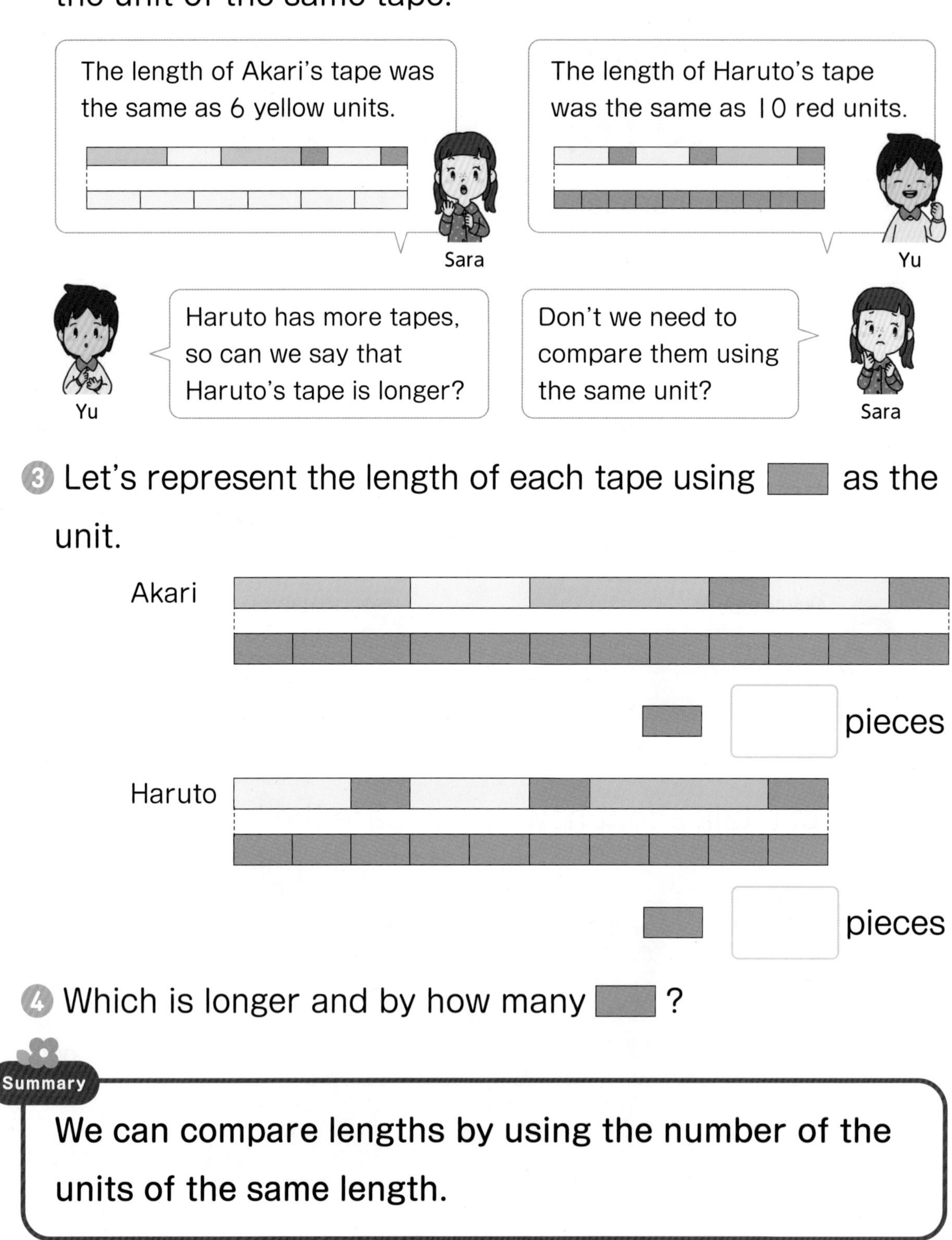

The length of Akari's tape was the same as 6 yellow units.
Sara
The length of Haruto's tape was the same as 10 red units.
Yu
Haruto has more tapes, so can we say that Haruto's tape is longer?
Yu
Don't we need to compare them using the same unit?
Sara

❸ Let's represent the length of each tape using ▭ as the unit.

Akari

▭ ☐ pieces

Haruto

▭ ☐ pieces

❹ Which is longer and by how many ▭ ?

Summary

We can compare lengths by using the number of the units of the same length.

? How can we measure the length of various objects?

2

Let's examine the width of the mathematics textbook by finding out how many ▭ on page 165 represents the width.

\ Want to explore /

Purpose How can we examine the length of various objects?

Akari

What shall I do when I don't have a tape?

Yu

1. We put the textbook on a grid paper. Let's compare the length and the width of the book.

2. Let's cut the grid paper to make a tool for measuring the length of various objects.

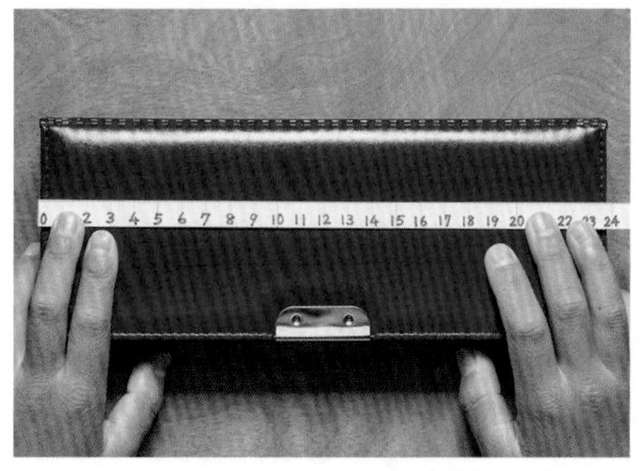

Sara

Summary We can measure the length of various objects by using the number of units of the same length.

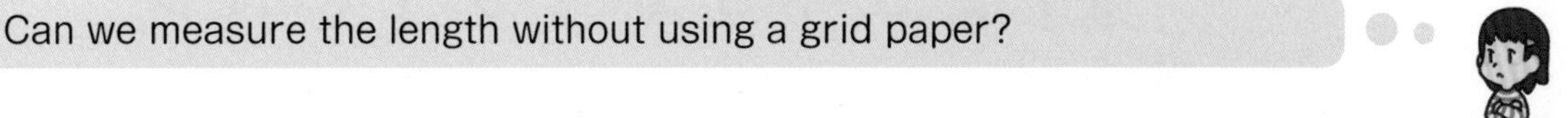

Can we measure the length without using a grid paper?

2 How to represent lengths

Let's measure the width of a postcard using the scales on the grid paper.

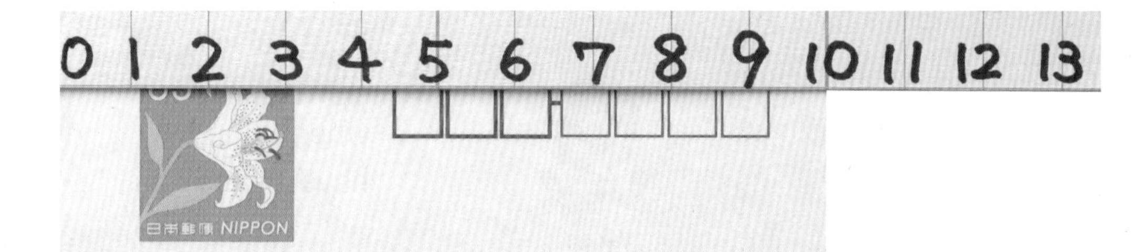

There is a **unit** called **centimeter** to measure length.
The length of one scale on the grid paper is written as **1 cm**, and is read as "1 centimeter."

1 cm

3 4 5 6 7

1 cm 1 cm

The width of a postcard is ☐ cm.

Many countries use cm as the unit of length.

Want to represent

? **Purpose** **Can we represent the length of various objects using cm?**

1 What is the length of the pencil in cm?

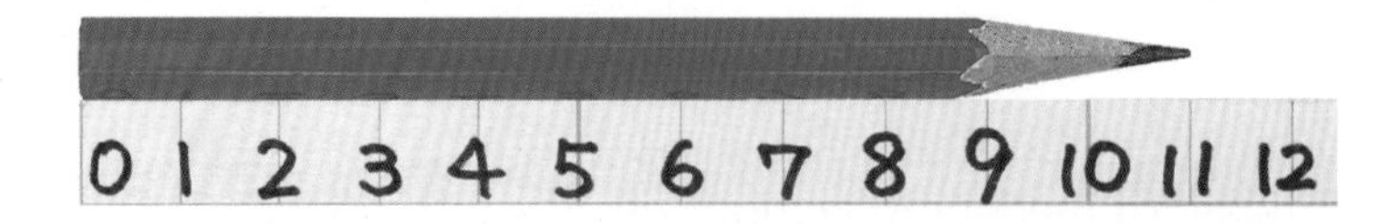

Summary

We can represent length by counting how many units of 1cm there are.

2 Which is the correct way to measure? Talk with your friends about the correct way to measure.

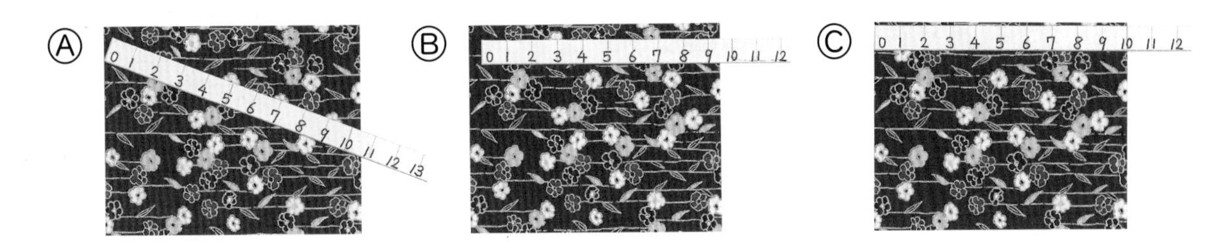

3 Let's measure the length of the tapes and the lines below.

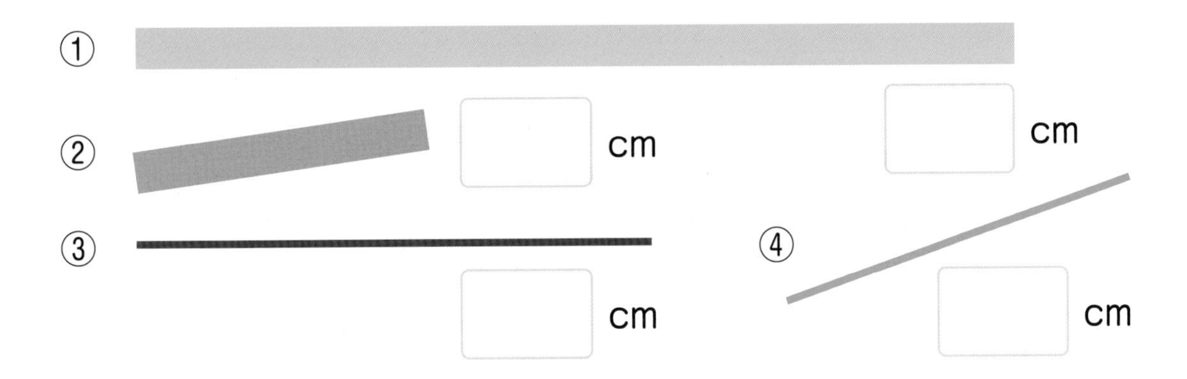

4 Using a 1cm scale, let's look for objects which are about 10cm long.

? How can we represent the length which is shorter than 1cm?

2 How can you represent the length of this stick?

How can we read the length represented between the units?

It would be better if there are smaller units.

\ Want to know /

Can we represent the length which is shorter than 1 cm?

If you use a **ruler**, you can measure the length which is shorter than 1 cm.

① The length of the stick is a little longer than 7 cm. How many smaller units are there beyond 7 cm?

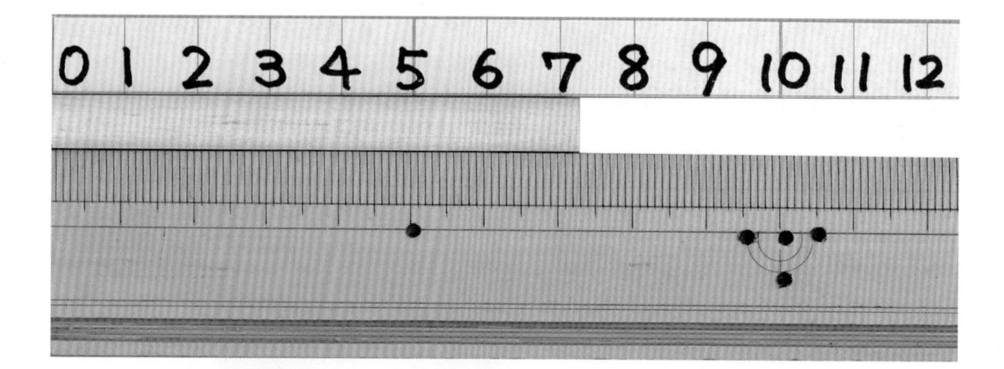

② How many smaller units is 1 cm divided into?

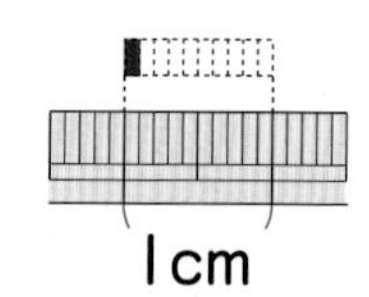

When the length of 1cm is equally divided into 10 parts, the unit of one part is written as 1mm, and is read as "1 **millimeter**."

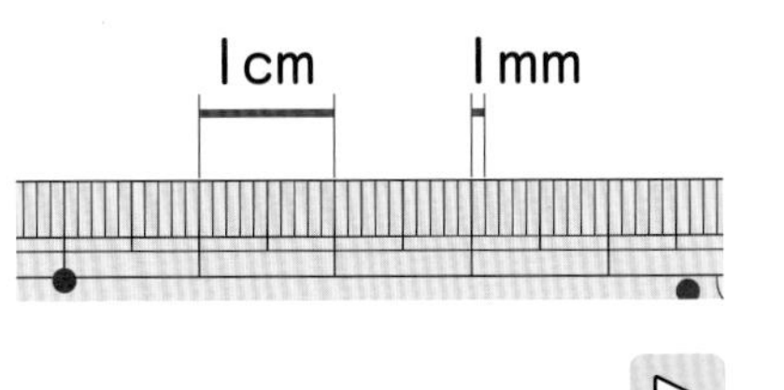

mm is also a unit used to measure length.

1cm = 10mm

The length of the stick is 7cm3mm. This is read as "7 centimeters and 3 millimeters."

1 From the left end of the ruler, what are the lengths up to ①, ②, ③, and ④ in cm and mm?

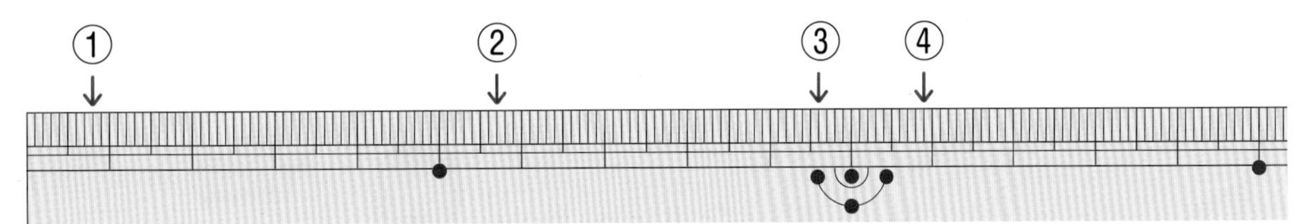

2 Let's measure the lengths of the lines.

①

②

③

④

Summary

We can represent length by setting 1mm as one unit and counting the number of it.

Way to see and think

? 1cm and 10mm represents the same length. Can we represent the length of other objects in two ways?

3 Let's measure the length of the side of an eraser.

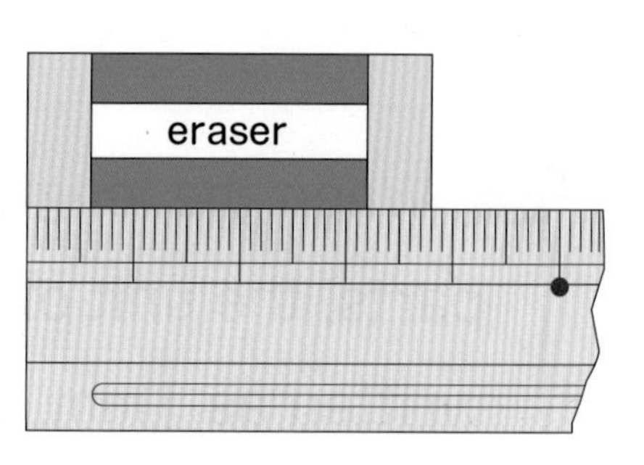

\ Want to represent /

Purpose Let's think about various ways to represent the same length.

① How many cm and mm is it?

② How many mm is it?

Ⓐ 3 cm = [ ] mm, so if we add 8 mm to it, we will get [ ] mm.

3 cm 8 mm = [ ] mm

Ⓑ There are 38 of 1 mm units, so it will be [ ] mm.

cm	mm
3	8

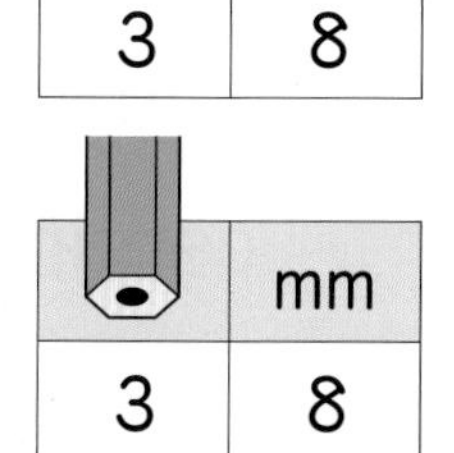

	mm
3	8

If we hide the "cm" unit, we can read it as "38 mm".

1 How many mm is 5cm? Also, how many cm and mm is 44mm?

Summary We can represent the same length in two different ways using cm and mm.

How to draw a straight line ▷

(1) Mark a point.

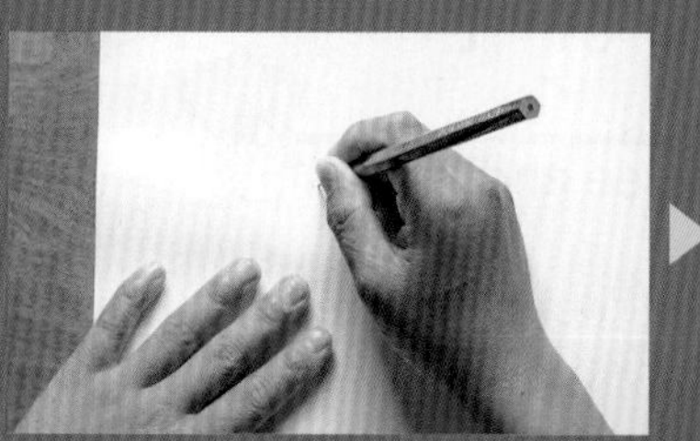

(2) Place one end of the ruler at the point.

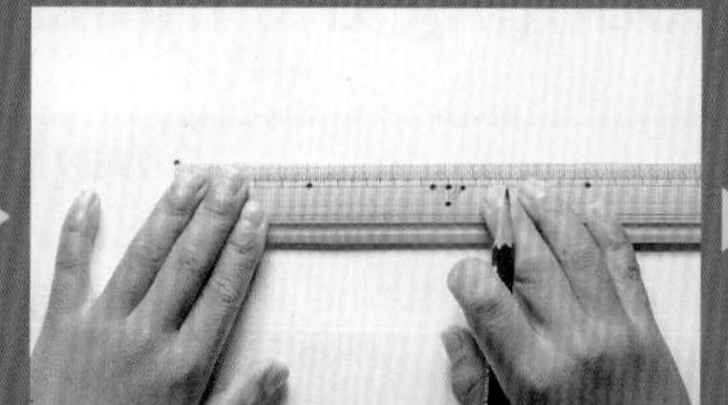

(3) Mark another point that is 8cm from the first point.

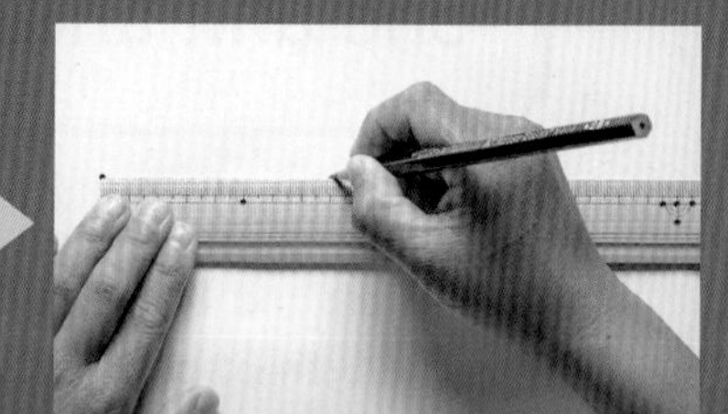

2 Let's fill in each ☐ with a number.

① 9cm = ☐ mm　② 7cm 9mm = ☐ mm

③ 80mm = ☐ cm　④ 62mm = ☐ cm ☐ mm

3 Which is longer, Ⓐ or Ⓑ?

① Ⓐ 6cm 3mm　Ⓑ 5cm 8mm

② Ⓐ 70mm　Ⓑ 7cm 4mm

How to draw a straight line →

\ Want to try /

Purpose How to draw a straight line is shown below.

Haruto

4 Let's draw lines that don't curve for the given lengths with a ruler.

❶ 8cm　❷ 11cm5mm　❸ 15cm8mm

A line that does not curve is called a **straight line**.

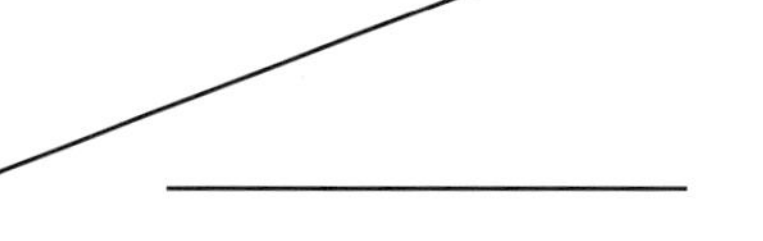

(4) Adjust the ruler to connect the two points.

(5) Draw the line by holding the ruler firmly.

Seen from the side

3 Calculating Length

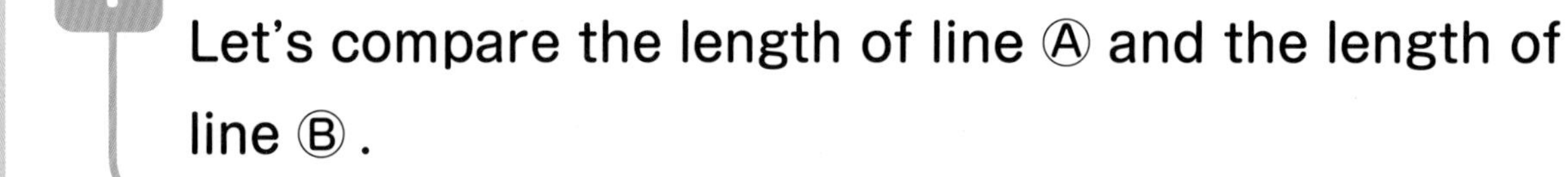

1 Let's compare the length of line Ⓐ and the length of line Ⓑ .

Because the lines are not straight, we cannot measure them.

Haruto

Can we add or subtract lengths?

Akari

\ Want to think /

Purpose How can we calculate lengths?

Way to see and think

They both align the lengths with the same unit.

① What is the length of line Ⓐ ?

Sara's idea

I used mm as a unit.

4cm2mm is ☐ mm.

3cm6mm is ☐ mm.

☐ mm+ ☐ mm = ☐ mm

Yu's idea

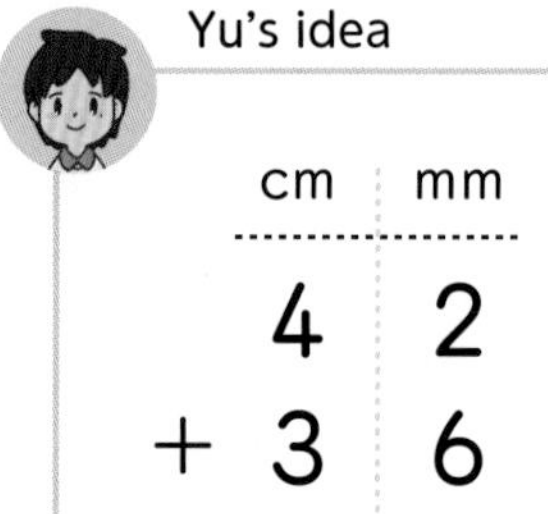

	cm	mm
	4	2
+	3	6
	☐	☐

We can calculate lengths in a similar way with vertical form.

Sara

☐ cm ☐ mm + ☐ cm ☐ mm = ☐ cm ☐ mm

② What is the length of line Ⓑ?

☐ cm ☐ mm + ☐ cm ☐ mm = ☐ cm ☐ mm

③ What is the difference between the length of line Ⓐ and the length of line Ⓑ?

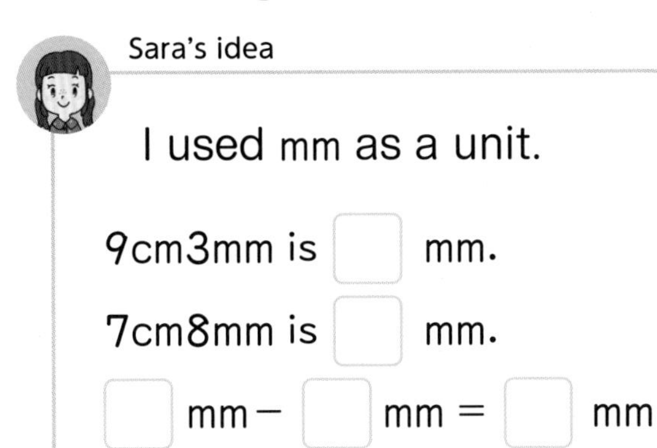

Sara's idea
I used mm as a unit.
9cm3mm is ☐ mm.
7cm8mm is ☐ mm.
☐ mm − ☐ mm = ☐ mm

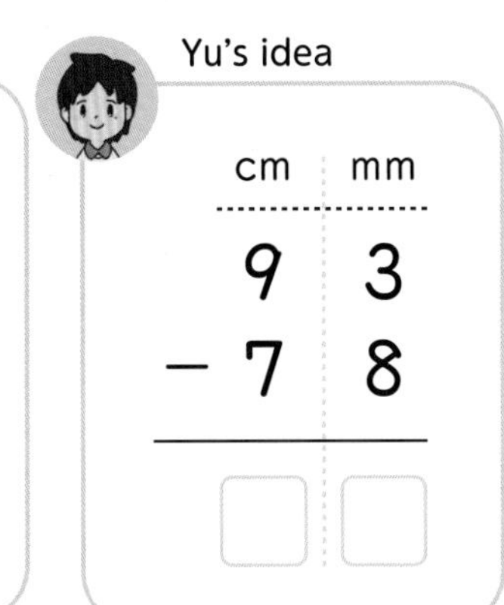

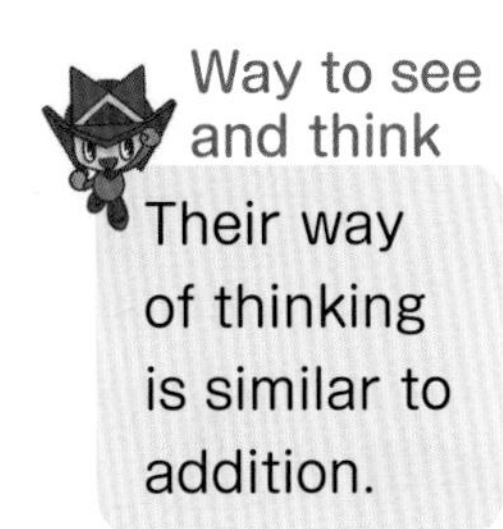

Way to see and think
Their way of thinking is similar to addition.

☐ cm ☐ mm − ☐ cm ☐ mm = ☐ cm ☐ mm

1 Let's think about how to calculate the following lengths.

① 7cm8mm + 2mm ② 6cm4mm − 4mm

Summary

We can calculate lengths by adding or subtracting numbers in the same units.

2 Let's calculate the following lengths.

① 12cm + 25cm ② 3mm + 8cm5mm

③ 3cm5mm + 4cm7mm ④ 28cm − 16cm

⑤ 18cm9mm − 6mm ⑥ 7cm6mm − 2cm9mm

CAN What can you do?

☐ We can measure the length with a ruler. → pp.70 ~ 71

1 From the left end of the ruler, what are the lengths up to ①, ②, ③ and ④ in cm and mm?

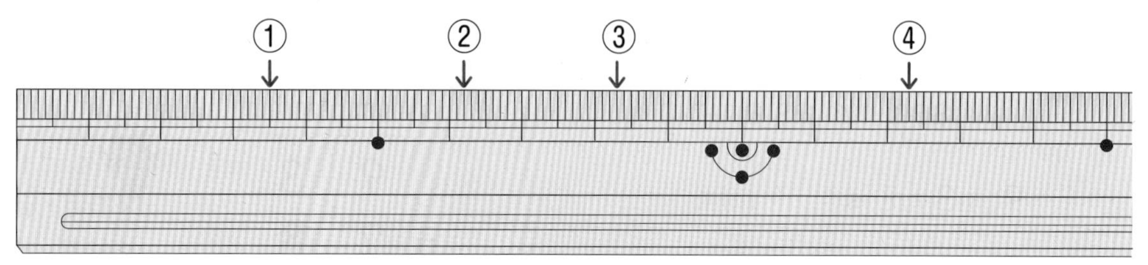

☐ We understand the conversion of the units of length. → p.72

2 Let's fill in the ☐ with numbers.

① 4cm= ☐ mm　　② 2cm5mm= ☐ mm

③ 60mm= ☐ cm　　④ 99mm= ☐ cm ☐ mm

☐ We can compare lengths. → pp.72 ~ 73

3 Which is longer, Ⓐ or Ⓑ?

① Ⓐ 7cm2mm　　Ⓑ 6cm9mm

② Ⓐ 11cm3mm　　Ⓑ 125mm

☐ We can calculate length. → pp.74 ~ 75

4 Let's calculate the following.

① 3cm4mm + 2cm3mm　　② 8cm9mm − 6cm5mm

Supplementary Problems → p.154

Which "Way to See and Think Monsters" did you find in 6 Length (1)"?

I found "Unit" when I represent length in numbers.

I found other monsters too!

Utilize Usefulness and Efficiency of Learning

1 Let's fill in each ☐ with the unit of length.

① The thickness of the notebook : 3 ☐

② The width of the desk : 60 ☐

2 There are two straight lines, line Ⓐ and line Ⓑ.

Ⓐ

Ⓑ

① Which is longer, line Ⓐ or line Ⓑ? By how much?

② What is the length when you connect line Ⓐ and line Ⓑ?

3 Read the rules and draw a line from the start to the treasure.

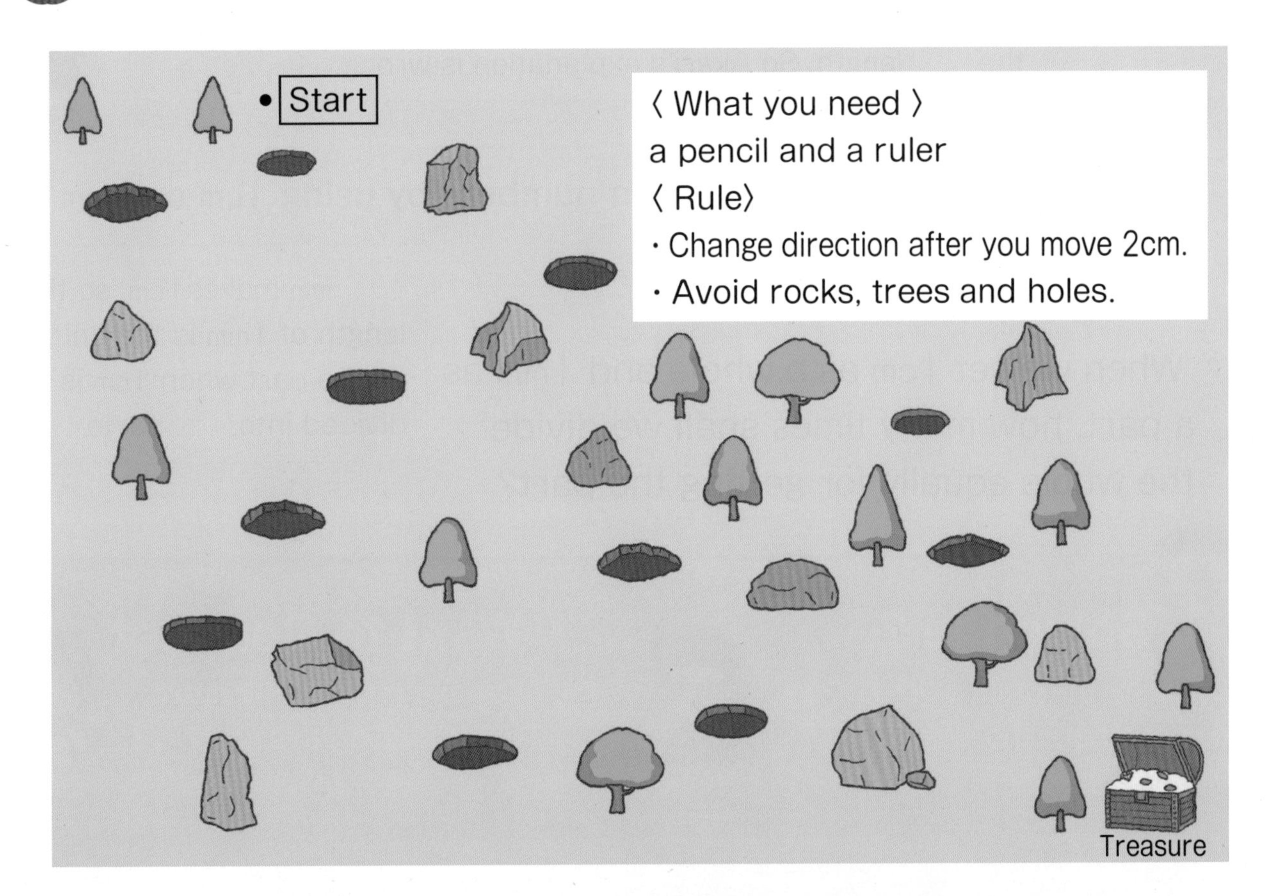

Start
〈 What you need 〉
a pencil and a ruler
〈 Rule〉
・Change direction after you move 2cm.
・Avoid rocks, trees and holes.
Treasure

With the Way to see and think Monsters...

Let's Reflect!

Let's reflect on which monster you found in "6 Length (1)."

Align

We could compare lengths by aligning the length of the unit.

① Two children are comparing the lengths of various objects as shown on the right. Akari is explaining the situation as follows. Is it correct?

Akari

The width of the pencil case is the same as 5 erasers. The length of the desk is the same as 3 pencils. So the width of the pencil case is longer than the length of the desk.

To measure length, we need to use the number of units of the ☐ length. So Akari's explanation is wrong.

Sara

Unit

We could represent length in numbers by using 1 cm or 1 mm as 1 unit.

② When we set 1 cm as a whole and 1 mm as a part, how many times shall we divide the whole equally for getting the part?

☐ mm makes 1 cm, so the length of 1 mm is the unit of one part when 1 cm is divided into ☐ parts.

Yu

Solve the ?

By using the units of length, we can compare the length of various objects.

Haruto

Want to Connect

Can we represent longer objects using this way?

Yu

Addition and Subtraction (1)

Let's think about how to calculate by using diagrams.

1 There are 12 red marbles and 14 blue marbles.
Let's think about the following.

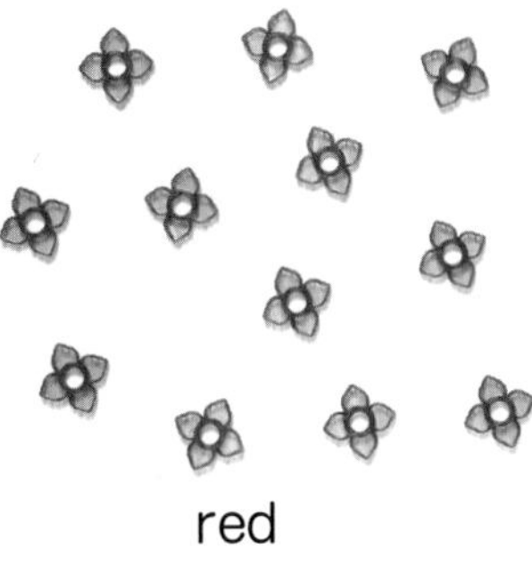

red

blue

① How many marbles are there in total?
Let's represent this situation by using a diagram.

Akari's diagram

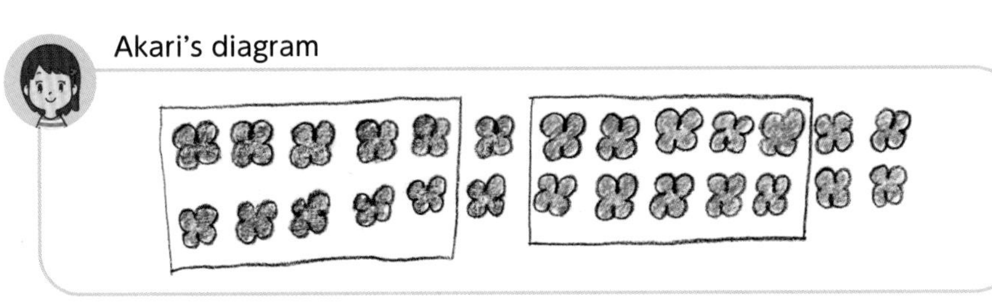

Haruto's diagram

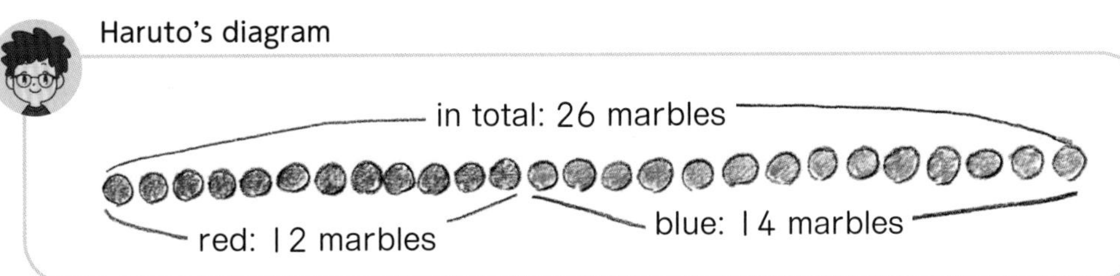

Sara's diagram

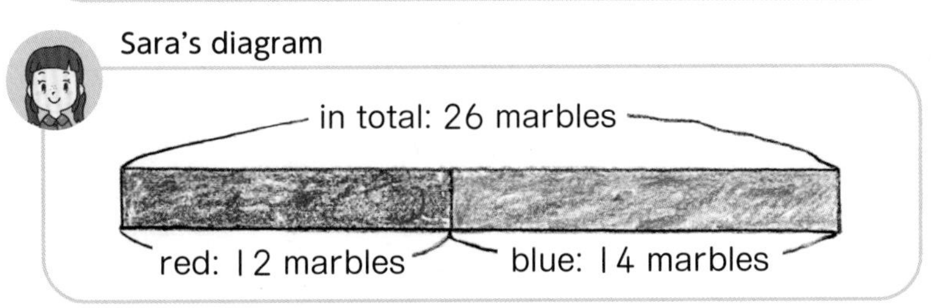

All three diagrams represent the number of marbles.

Yu

There is a diagram which may be useful even when the number gets large.

Akari

② Let's talk about the good points of each diagram.

＼ Want to represent ／

? Purpose Can we represent the problem using a diagram?

1 In the 2nd term, there were 29 children in Hitomi's class. 3 new children joined the class in the 3rd term. There are 32 children in total. Let's represent this problem in order from (1) to (3) with diagrams. ▷

(1) There were 29 children in the 2nd term.

2nd term: ☐ children

We can use the approximate length.

(2) 3 children joined.

2nd term: ☐ children ☐ children joined

(3) There are 32 children in total.

in total: ☐ children

2nd term: ☐ children ☐ children joined

! Summary

We can understand the situation easier by using diagrams and representing the problem in order.

2 There are 21 red pencils and 23 blue pencils. There are 44 pencils in total. Let's represent this situation by using a diagram.

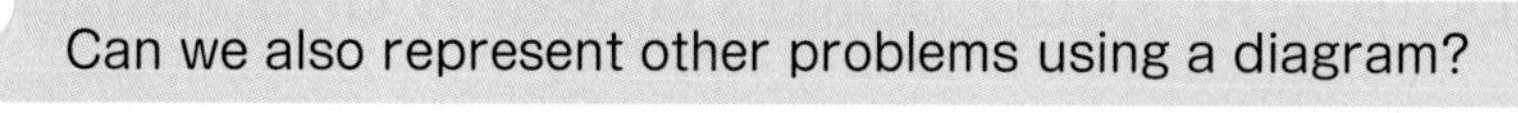

Can we also represent other problems using a diagram?

2

There was a 23 cm tape. You used 17 cm of it. How many cm of the tape are left?

\ Want to solve /

? **Purpose** How can we calculate to find out the answer?

❶ Let's represent this problem in order from (1) to (3) with diagrams. ▷

(1) There was a 23 cm tape.

At the start: 23cm

(2) 17 cm of it was used.

At the start: 23cm

Used: 17cm

(3) How many cm are left?

At the start: 23cm

Used: 17cm　Left: □cm

Let's think about what should be represented as □.

❷ Let's write a math expression and the answer.

! **Summary**

We can find out how to calculate by using diagrams.

▶1 There were 50 sheets of colored paper. You used 16 sheets of them. How many sheets are left?

? Can we represent any problem with a diagram?

3

Aoi picked up 18 acorns. Her teacher picked up 22 acorns. What is the difference of the number of acorns?

❶ Let's represent this problem in order from (1) to (2) with diagrams. ▷

Way to see and think

Let's think about what should be represented as □.

(1) Aoi picked up 18 acorns.

18 acorns

Aoi

(2) Her teacher picked up 22 acorns.

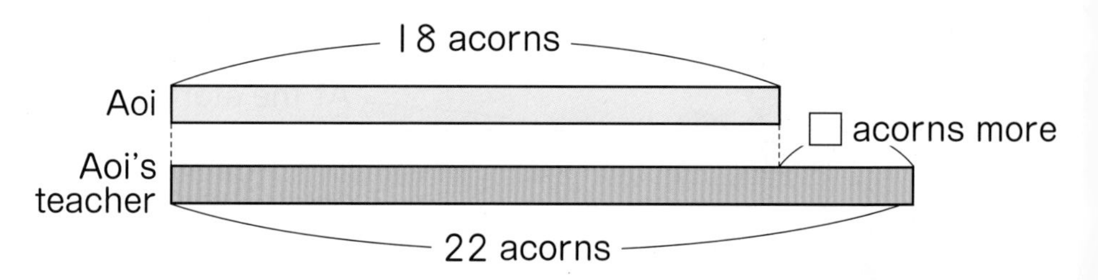

Sara: There are two tapes.

Akari: What is the difference between the diagram with one tape?

\ Want to solve /

? **Purpose** **In what situation is it easier by drawing a diagram with two tapes?**

❷ Let's write a math expression and the answer.

1 Saki picked up 31 empty cans. Sota said he picked up 5 more cans than Saki. How many cans did Sota pick up?

① Let's represent this problem with diagrams.

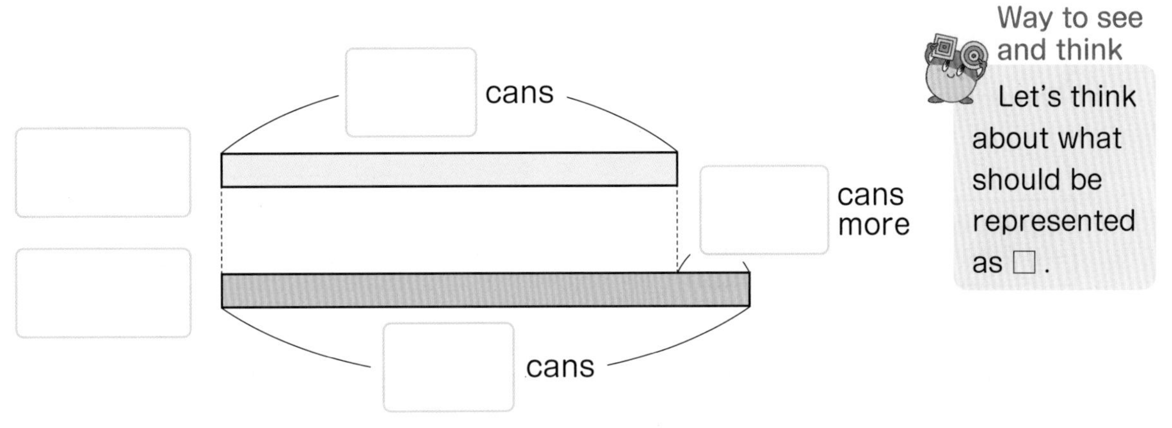

Way to see and think

Let's think about what should be represented as □.

② Let's write a math expression and the answer.

Summary

When we want to compare the numbers and find out the difference, it is easier to compare by drawing a diagram with two tapes.

2 There are 32 red tulips. There are 8 fewer white tulips than the red tulips. How many white tulips are there?

How can we draw a diagram for this situation?

4

We took a group photo.
There were 8 chairs, where 1 child sat on each, and 13 children stood.
How many children were there in the photo?

\ Want to think /

Purpose Can we compare different objects?

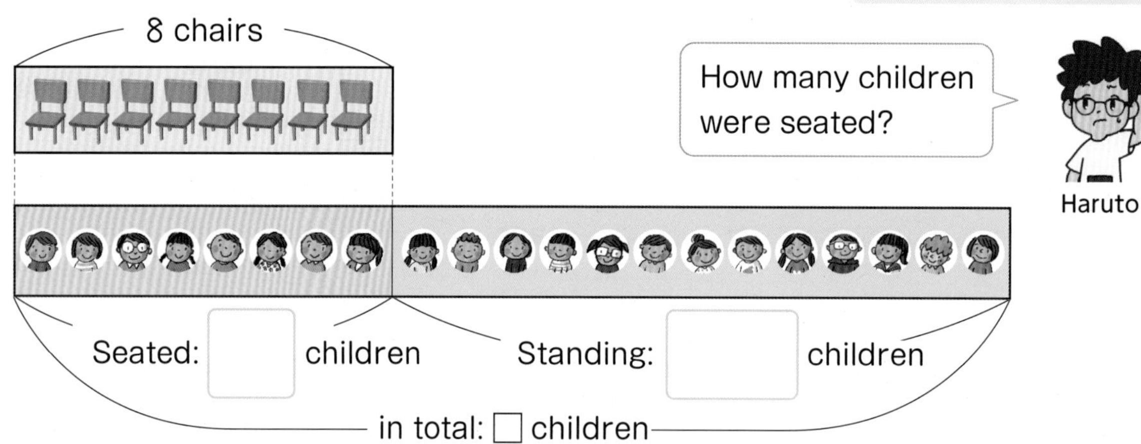

1 There are 15 cakes. When there are 6 plates with one cake each, how many cakes are left?

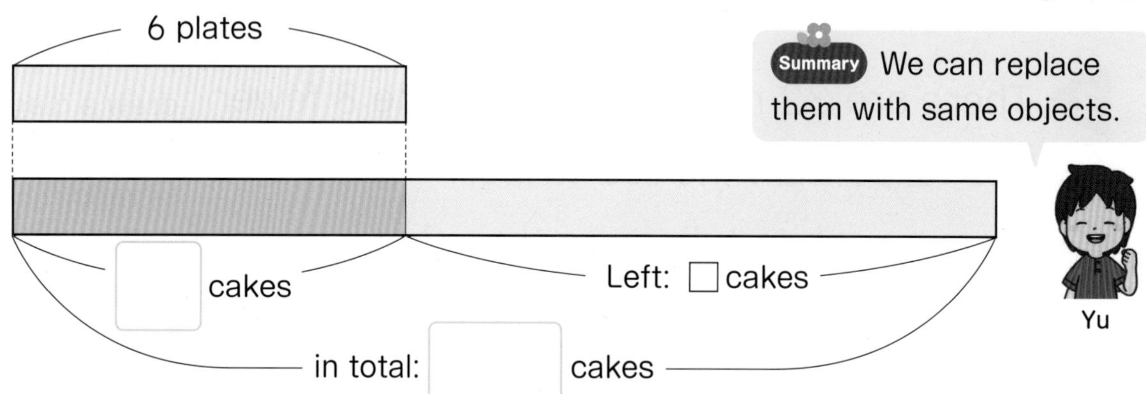

Summary We can replace them with same objects.

CAN What can you do?

☐ We understand the relationship between the situation and the diagram. → p.81

1 A diagram that matches the following situation is shown below.

Let's fill in each ( ) with a phrase from ☐ and complete the diagram. Then let's find out the answer.

There are 31 killifish in the left aquarium and 18 killifish in the right aquarium. How many in total?	Right aquarium: 18 killifish Total: ☐ killifish Left aquarium: 31 killifish

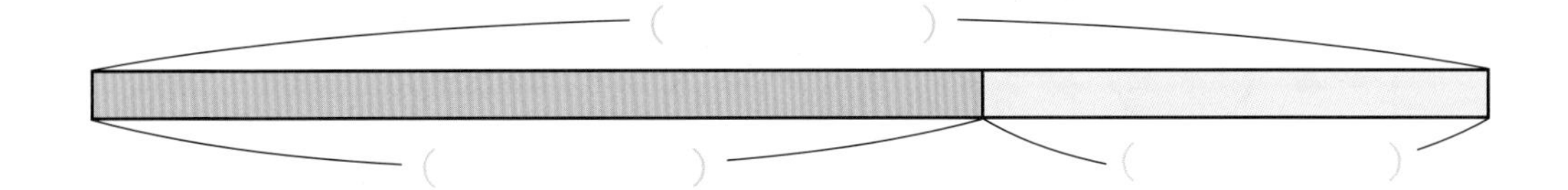

☐ We can make a math expression to find out the answer. → pp.82～83

2 Nanami folded 18 cranes. Yui folded 23 cranes. What is the difference of the number of cranes?

① Let's complete the diagram on the right.

② Let's write a math expression and the answer.

Nanami ☐ cranes

Difference: ☐ cranes

Yui ☐ cranes

Supplementary Problems → p.155

Which "Way to See and Think Monsters" did you find in "7 Addition and Subtraction (1)"?

I found "Other Way" when I was representing the problem with diagrams.

I found other monsters too!

8

Numbers up to 1000

Let's explore how to represent numbers and their structure.

So many chicks! How many are there?
Akari
Can you find an easier way to count?

1 Numbers larger than 100

1 How many chicks are there in total?

I circled the chicks in sets of 10.

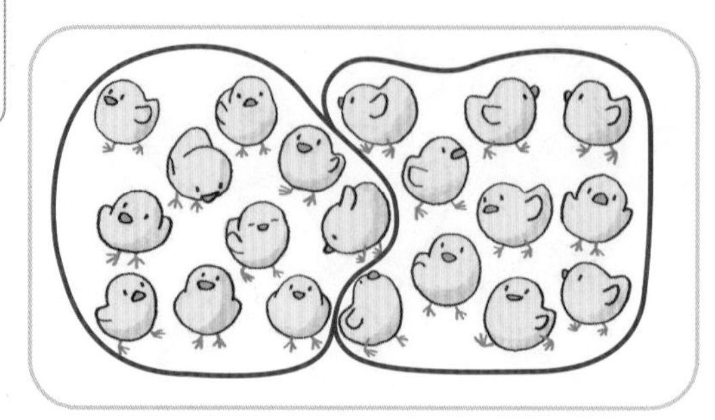

We can find a lot of sets.

Way to see and think

It's easier to count by setting them into units.

1. Let's put ▢ in boxes of 10 blocks. ▷

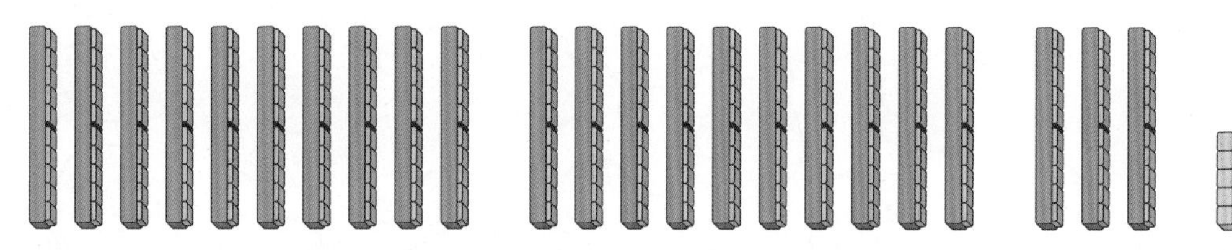

There are ☐ boxes of 10 blocks and ☐ single blocks.

Way to see and think

It's easier to replace the chicks with ▢.

10 sets of 10 make 100...

Sets of 100...

Want to represent

? **Purpose** How can we represent the numbers larger than 100?

2. 10 boxes of 10 make 100. How many boxes of 100 can we make? ▷

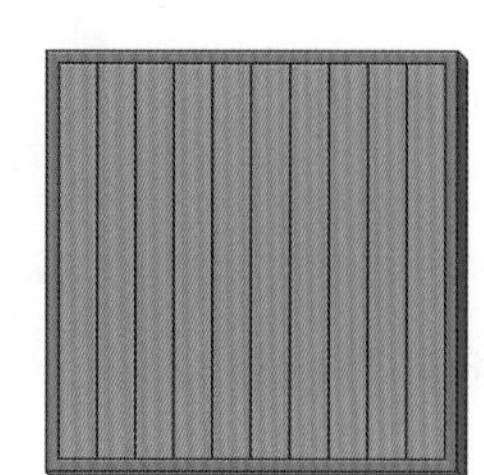

If we make a box of 100, how many boxes of 10 and singles will be left?

Yu

Two sets of 100 is **two hundred**. Two hundred, thirty, and five is represented as **two hundred thirty-five**, and is written as 235. The position of 2 in 235 is called the **hundreds place**.

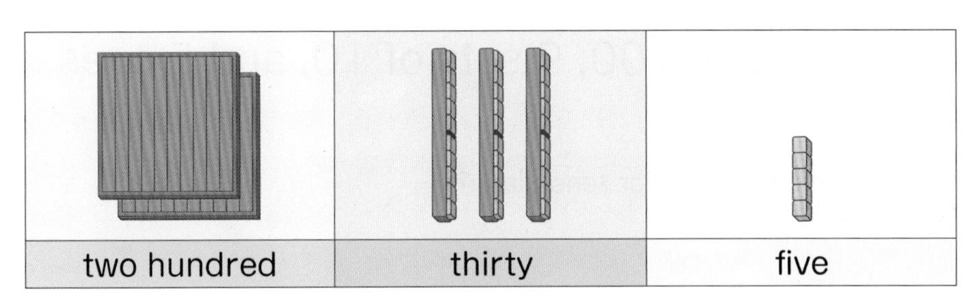

Hundreds Place	Tens Place	Ones Place
2	3	5

It's the sum of 2 sets of 100, 3 sets of 10, and 5 ones.

1 How many ▢ are there in total?

①

Hundreds	Tens	Ones

Hundreds Place	Tens Place	Ones Place

②

Hundreds	Tens	Ones

Hundreds Place	Tens Place	Ones Place

Summary

We can represent numbers larger than 100 by using the hundreds place and grouping the numbers of each place.

2 Let's read the following numbers.

① 379　　② 516　　③ 847　　④ 136

3 Let's write the following numbers in digits.

① The number when seven hundred, thirty and four are added.

② The number when one hundred, fifty, and seven are added.

③ The sum of 4 sets of 100, 9 sets of 10, and 5 ones.

? What if we didn't have anything for tens place?

Representing zero →

2 How many pencils are there?

Akari

We don't have any set of 10.

We have sets of 100 and singles only...

Haruto

\ Want to represent /

? **Purpose** **If we don't have anything for tens place, how can we represent?**

Hundreds	Tens	Ones

Hundreds Place	Tens Place	Ones Place

☐ sets of 100, ☐ sets of 10, ☐ ones

The number of pencils is written as 307 and read as **three hundred seven**.

1 How many ▢ are there in total?

①

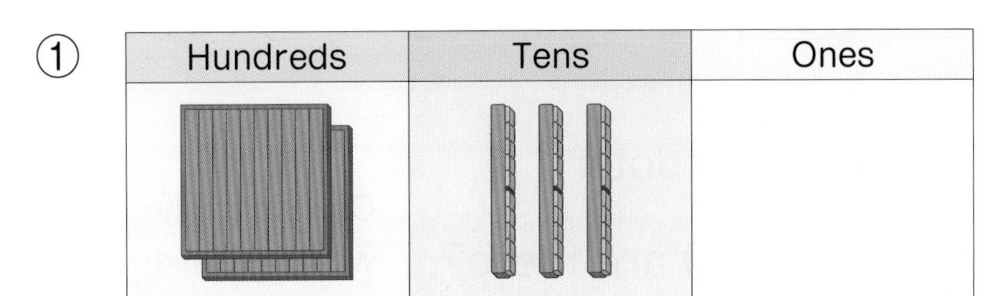

Hundreds	Tens	Ones

Hundreds Place	Tens Place	Ones Place

the sum of 200 and 30

②

Hundreds	Tens	Ones

Hundreds Place	Tens Place	Ones Place

the sum of 3 sets of 100

Summary

When we don't have anything for the tens place or the ones place, we represent it with 0 for that place.

2 Let's read the following numbers.

① 820 ② 160 ③ 408 ④ 900

3 Let's write the following numbers.

① seven hundred forty ② one hundred twenty

③ five hundred eight ④ six hundred

4 Let's fill in each ▢ with a number.

① The sum of 5 sets of 100 and 6 ones is ▢.

② 860 is the sum of ▢ sets of 100 and ▢ sets of 10.

? What will happen if we collect 10 sets of 100?

3 How many ● are there in the diagram below?

❶ Circle the sets of 100 dots.

❷ How many sets of 100 are there?

Can we apply what we have learned before?
Sara

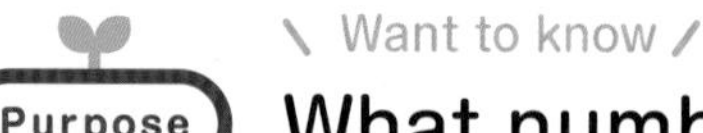

? **Purpose** What number does 10 sets of 100 make?

10 sets of 100 is **one thousand** and is written as 1000.

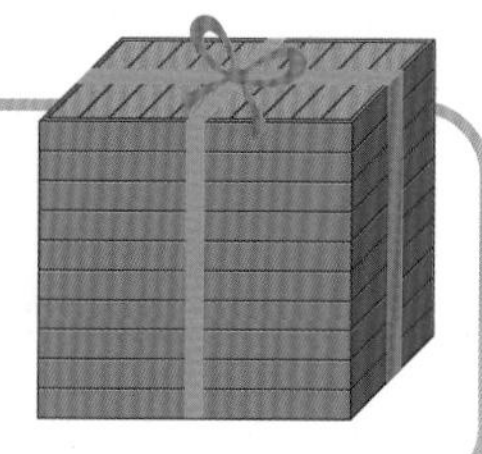

1 Let's write the following numbers by using the line of number below.

① What is the number that is added to 900 to get to 1000?

② What is the number that is 10 smaller than 1000?

③ What is the number that is 1 smaller than 1000?

Way to see and think

Let's think about what the smallest scale on the line represents.

! **Summary**

The number that is the sum of 10 sets of 100 will become an higher place as we learned.

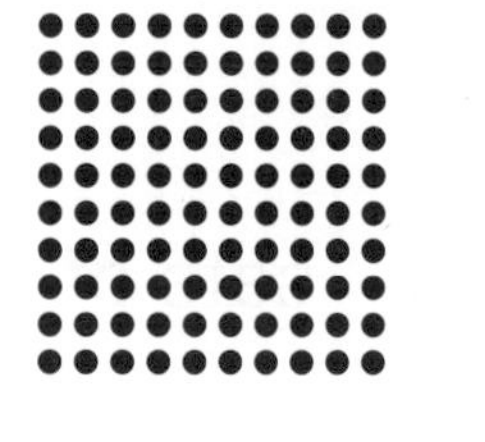 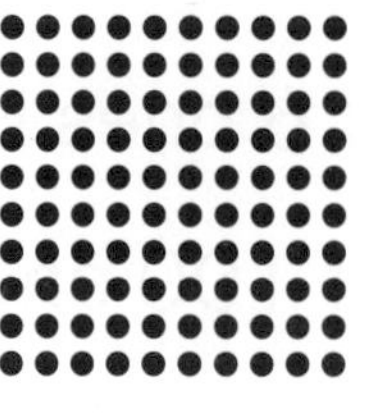   

0 100 200 300 400 500

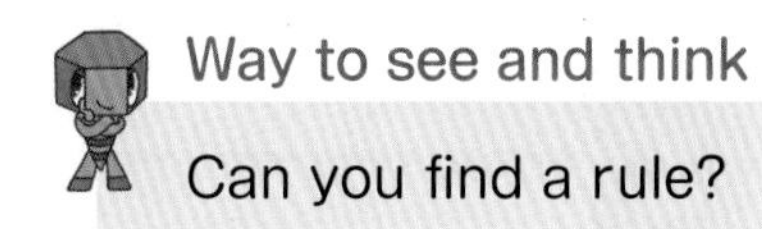

Way to see and think
Can you find a rule?

2 Let's fill in each ☐ with a number.

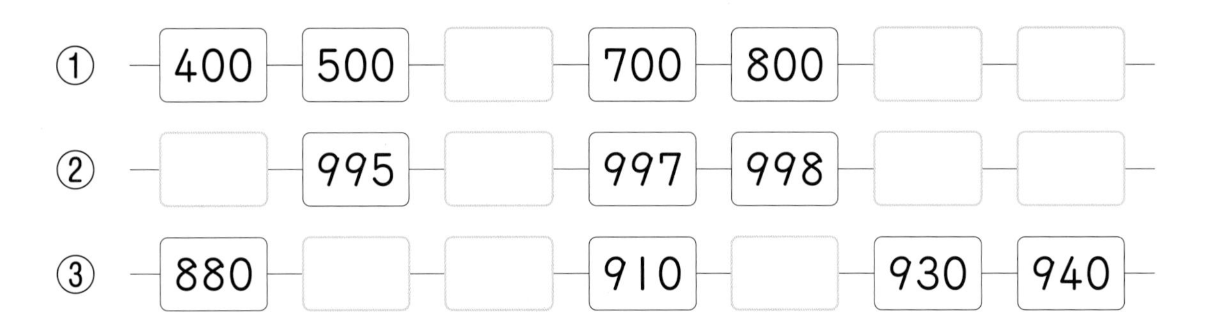

3 Let's write the numbers in the ☐ that each ↑ is pointing at.

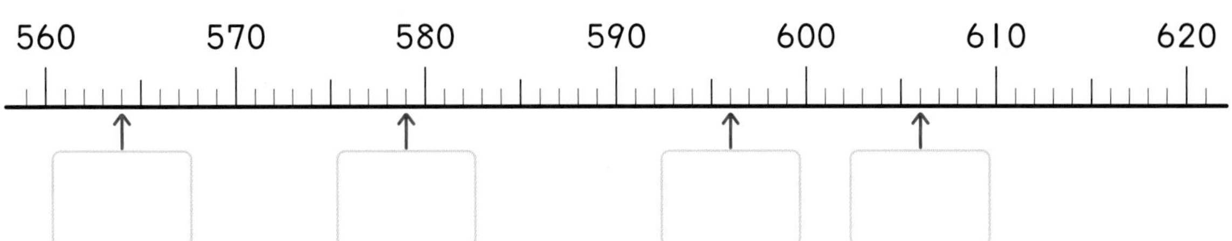

4 Let's draw ↑ at the scale on the line of number below that represents each number.

① 450 ② 680 ③ 990

5 Let's write the following numbers.

① The number that is 300 larger than 500.

② The number that is 200 smaller than 700.

? 10 sets of 10 make 100. What happens if we have 20 sets of 10, 30 sets of 10?

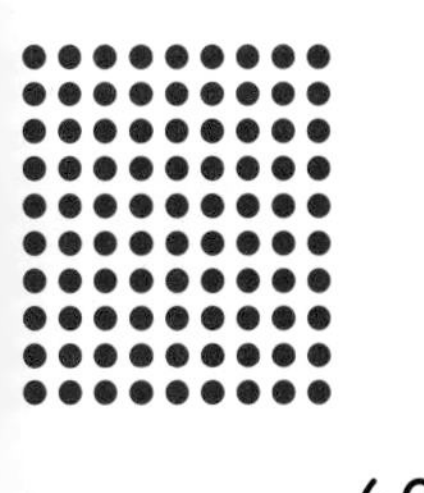    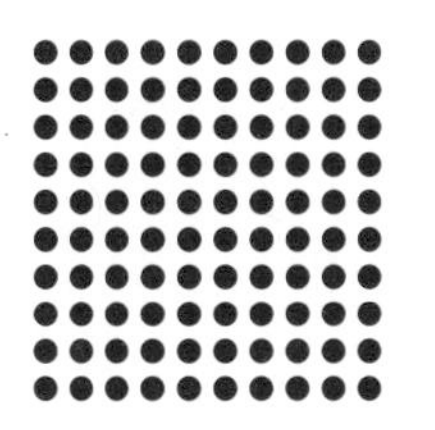

600 700 800 900 1000

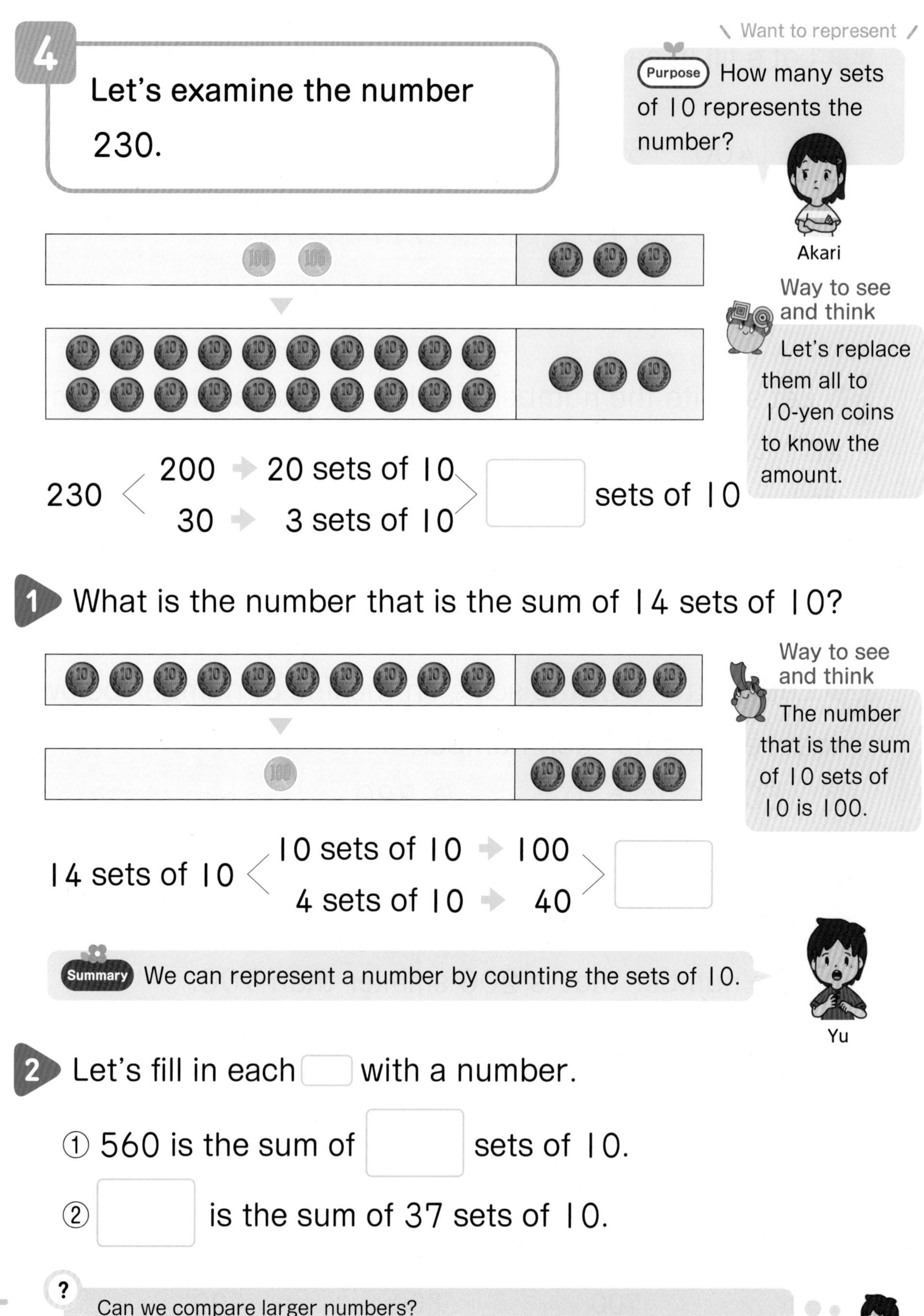

4 Let's examine the number 230.

Want to represent

Purpose How many sets of 10 represents the number?

Akari

Way to see and think: Let's replace them all to 10-yen coins to know the amount.

230 < 200 → 20 sets of 10, 30 → 3 sets of 10 > ☐ sets of 10

1 What is the number that is the sum of 14 sets of 10?

Way to see and think: The number that is the sum of 10 sets of 10 is 100.

14 sets of 10 < 10 sets of 10 → 100, 4 sets of 10 → 40 > ☐

Summary We can represent a number by counting the sets of 10.

Yu

2 Let's fill in each ☐ with a number.

① 560 is the sum of ☐ sets of 10.

② ☐ is the sum of 37 sets of 10.

? Can we compare larger numbers?

2 Small and Large Numbers

Comparing numbers ↓

1 The table on the right shows the numbers of plastic bottle caps that were collected by the Class 1 and Class 2. Let's compare the numbers of caps. Which class collected more caps?

Class 1	290 caps
Class 2	288 caps

＼ Want to compare ／

? **Purpose** How do we compare the size of large numbers?

① Let's compare the numbers by writing numbers in the table on the right.

	Hundreds Place	Tens Place	Ones Place
Class 1			
Class 2			

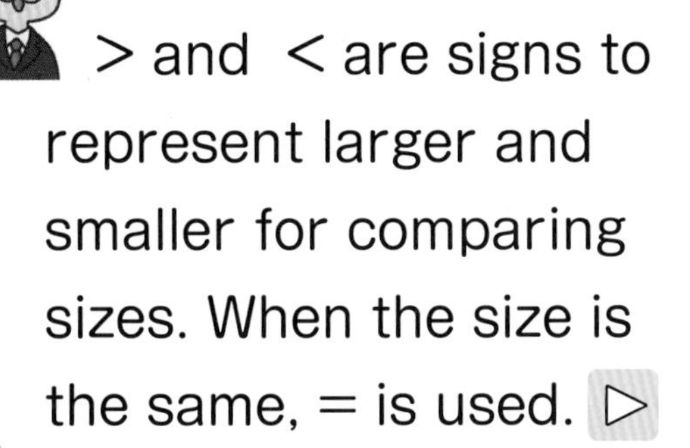

$>$ and $<$ are signs to represent larger and smaller for comparing sizes. When the size is the same, $=$ is used. ▷

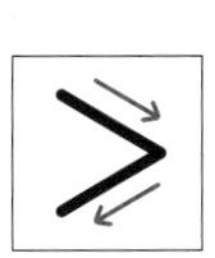

$4 > 2$	$3 = 3$	$2 < 4$
4 is larger than 2.	3 is the same size as 3.	2 is smaller than 4.

1 Let's write $>$ or $<$ in each □.

① 238 □ 253　　② 769 □ 764

! **Summary**

We can compare large numbers by using a table with aligned places and line of number.

? Can we do addition and subtraction using large numbers?

1

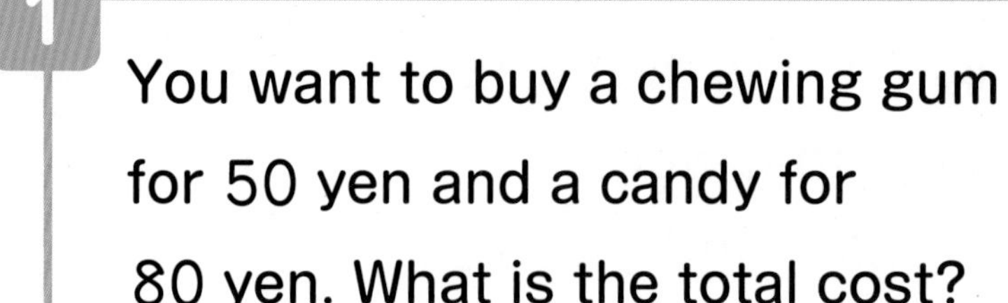

You want to buy a chewing gum for 50 yen and a candy for 80 yen. What is the total cost?

① Let's write a math expression.

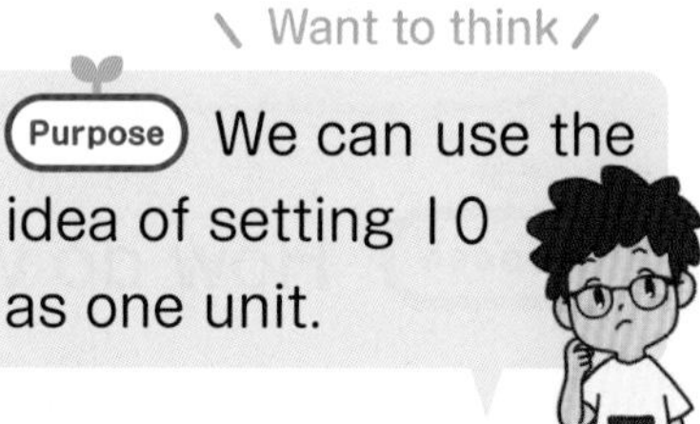

Want to think
Purpose We can use the idea of setting 10 as one unit.
Haruto

② Let's think about the way of calculation.

Way to see and think
Let's think about how many sets of ten...

1 You have 130 yen. If you buy a chocolate for 70 yen, how much will be left?

Math Sentence:

Way to see and think
Let's think about how many sets of 10 there are in 130.

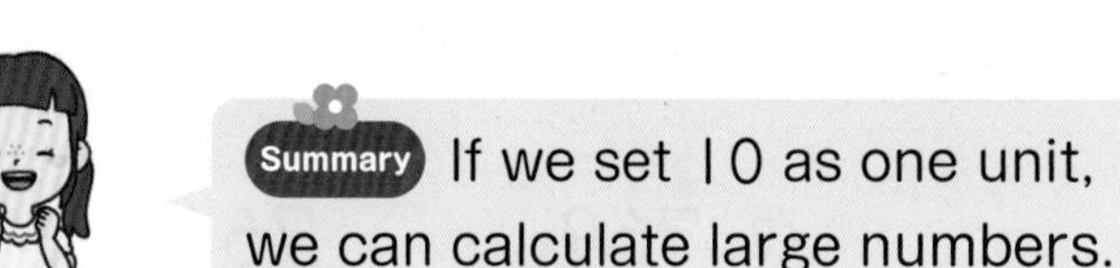

Summary If we set 10 as one unit, we can calculate large numbers.
Sara

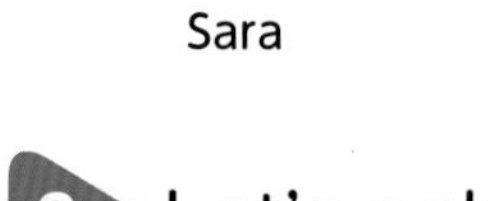

2 Let's calculate the following.

① 70 + 50　② 90 + 20　③ 60 + 60

④ 130 − 50　⑤ 140 − 80　⑥ 160 − 90

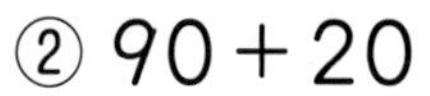

CAN What can you do?

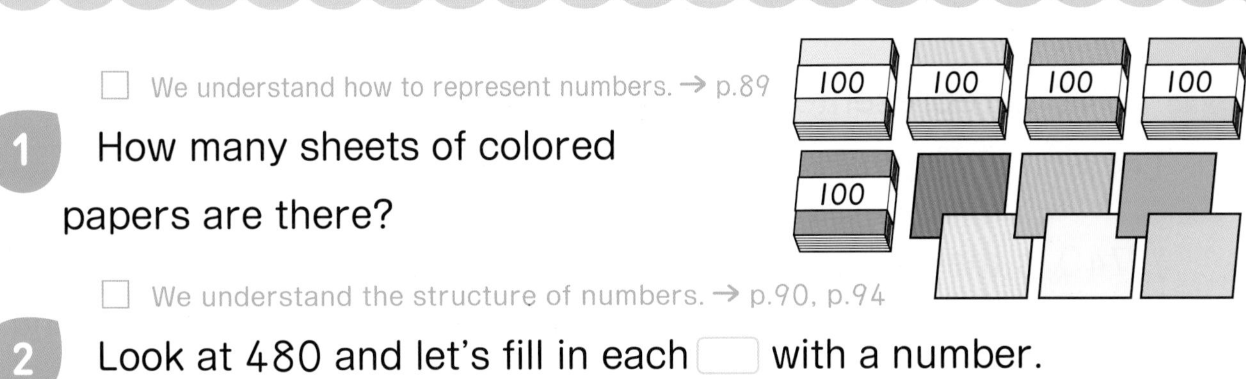

☐ We understand how to represent numbers. → p.89

1 How many sheets of colored papers are there?

☐ We understand the structure of numbers. → p.90, p.94

2 Look at 480 and let's fill in each ☐ with a number.

① 4 in the hundreds place means that 4 has a value of ☐.

② 480 is the sum of ☐ sets of 10.

☐ We understand the structure of numbers. → pp.89 ~ 91

3 Let's fill in each ☐ with numbers.

① The sum of five hundred, ten, and nine is ☐.

② The sum of 2 sets of 100 and 5 ones is ☐.

☐ We understand the sizes of 3-digit numbers. → p.95

4 Let's write > or < in each ☐.

① 718 ☐ 781　　② 555 ☐ 559　　③ 310 ☐ 301

☐ We can use sets of 10 to calculate. → p.96

5 Let's calculate the following.

① $40 + 90$　　② $70 + 70$　　③ $130 - 60$　　④ $150 - 80$

Supplementary Problems → p.156

Which "Way to see and Think Monsters" did you find in "8 Numbers up to 1000"?

I found "Unit" when I was exploring large numbers.

I found other monsters too!

Utilize Usefulness and Efficiency of Learning

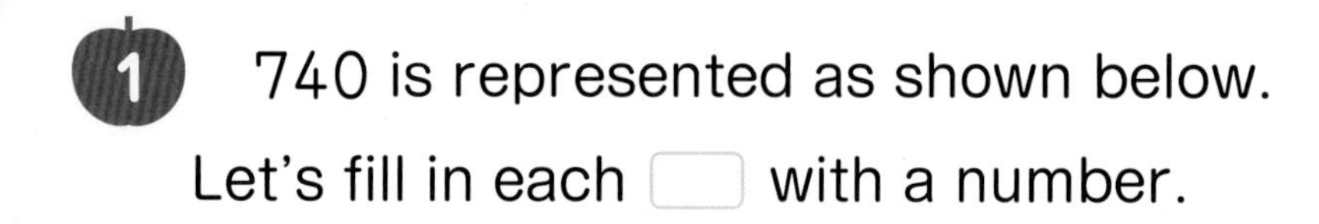

1 740 is represented as shown below.
Let's fill in each □ with a number.

740 is the number that is 60 smaller than □.	740 is the number when □ and 40 are added.	740 is the number that is the sum of □ sets of 10.

2 The number in the ones place of the White Team's score is still unknown. Takuto says that the Red Team has won.
Why does he say so?

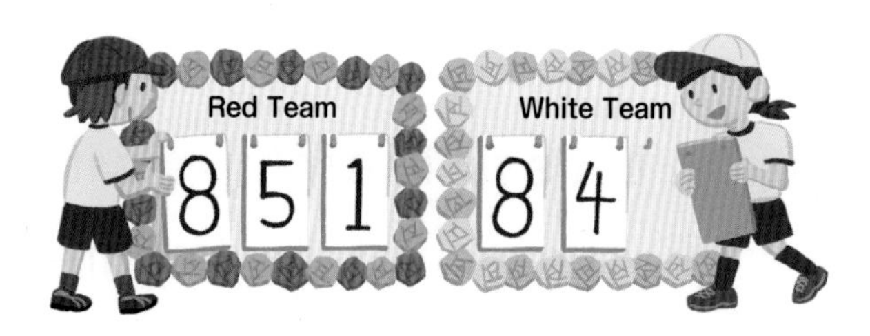

3 Yuko went shopping. In her wallet, there are two 100-yen coins. Additionally, there are total of 5 coins, including both 10-yen coins and 1-yen coins. Let's answer the following problem.

① 200 yen and how much more does Yuko have? Write down the numbers of 10-yen coins and 1-yen coins and think about various ways.

One 10-yen coin and four 1-yen coins : 14 yen
Two 10-yen coins and three 1-yen coins : 23 yen
□ 10-yen coins and □ 1-yen coins : □ yen
□ 10-yen coins and □ 1-yen coin : □ yen

If she had three 10 yen coins...

Akari

② Yuko paid 40 yen to buy a bubble gum, and she got change.
She has 8 coins in her wallet. Which coin did she pay, and which coin did she get change?

With the Way to see and think Monsters...

Let's Reflect!

Let's reflect on which monster you found in "8 Numbers up to 1000."

Unit

We could represent large numbers by considering the hundreds place, tens place, and ones place to be one unit each.
Larger numbers can be represented by how many sets of 10 it is consisted of, setting 10 as one unit.

① 628 is the sum of ☐ sets of 100, ☐ sets of 10, and ☐ ones.

Similar to what we have learned before, we thought about 3-digit numbers by separating each places.
Haruto

② 630 is the sum of ☐ sets of 10.

③ The number that is the sum of 29 sets of 10 is ☐.

④ Let's think about how to calculate $40 + 60$.

Ⓐ 40 is the sum of ☐ sets of 10.

Ⓑ 60 is the sum of ☐ sets of 10.

Ⓒ Ⓐ and Ⓑ is the sum of ☐ sets of 10.

Ⓓ $40 + 60 =$ ☐

Solve the ?

If we set 100, 10, and 1 as one unit, we can represent large numbers.

Want to Connect

What is the number which has 10 sets of 1000?

Find the ?

How many plastic bottles?

Let's think about how to calculate large numbers.

Addition and Subtraction of Large Numbers

Let's think about the meaning of calculation and how to calculate.

1 Addition of 3-digit answers

Addition with carrying to the hundreds place →

1 Akari and Yu participated in the cleaning the town. Akari's group picked up 74 plastic bottles. Yu's group picked up 65 plastic bottles.
How many bottles did they pick up in total?

About how many did they pick up?

① Let's write a math expression.

Way to see and think
Remember the addition of 2-digit numbers in vertical form.

② Let's think about how to calculate.

If we separate the tens place and the ones place...

If we use blocks...

74 is $70+4$, and 65 is $60+5$, so...

What do we do with the tens place if we use the vertical form?

Want to think

? Purpose **Can we also calculate addition of large numbers in vertical form?**

③ Let's explain how to add $74+65$ in vertical form.

	7	4
+	6	5

Addition algorithm for $74 + 65$ in vertical form

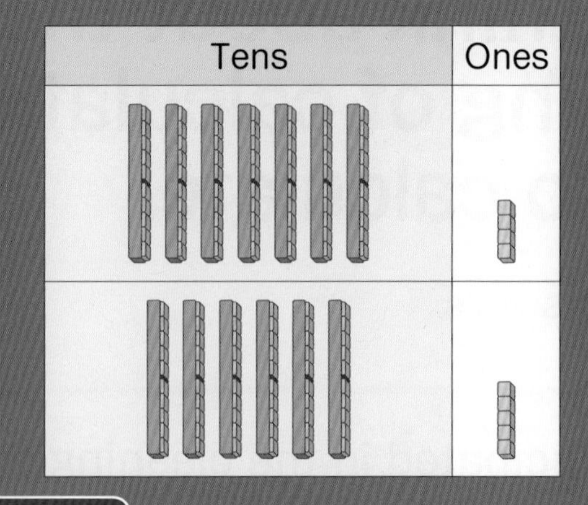

	7	4
+	6	5

Align the digits of the numbers according to their places.

↓

Ones Place

$4 + 5 = 9$

	7	4
+	6	5
		9

The number in the ones place is 9.

↓

Tens Place

$7 + 6 = 13$

	7	4
+	6	5
1	3	9

The tens place is 3. Carry 10 tens to the hundreds place as 1 hundred.

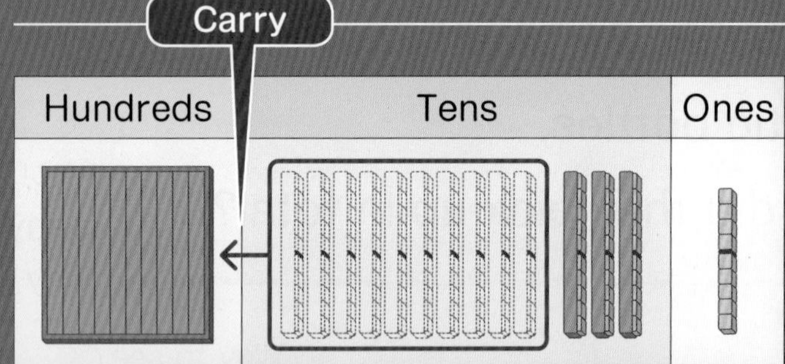

Math Sentence: $74 + 65 = 139$ Answer: 139 bottles

1 Let's think about how to add $93 + 36$ in vertical form.

Summary

We can add large numbers in vertical form by separating the digits of the numbers according to their places.

2 Let's calculate the following in vertical form.

① $63 + 71$ ② $67 + 80$ ③ $20 + 90$

? In addition when we have carrying from the ones place to the tens place, can we calculate in the same way as we have learned?

2

Let's explain how to add $74+58$ in vertical form.

	7	4
+	5	8

\ Want to think /

? **Purpose** What should we do when we calculate addition with carrying?

Way to see and think

Can addition with carrying be done in the same way as we have learned?

Addition algorithm for $74+58$ in vertical form

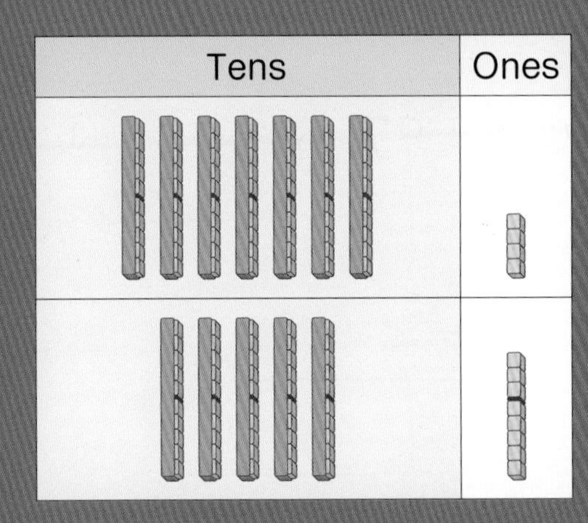

$$\begin{array}{r} 74 \\ +\ 58 \\ \hline \end{array}$$

Align the digits of the numbers according to their places.

↓

Ones Place

$4+8=12$

$$\begin{array}{r} 74 \\ +\ 58 \\ \hline {}^{1}\ \ \\ 2 \end{array}$$

The number in the ones place is 2. Carry 10 ones to the tens place as 1 ten.

↓

Tens Place

$7+5+1=13$

$$\begin{array}{r} 74 \\ +\ 58 \\ \hline {}^{1}\ \ \\ 132 \end{array}$$

The tens place is 3. Carry 10 tens to the hundreds place as 1 hundred.

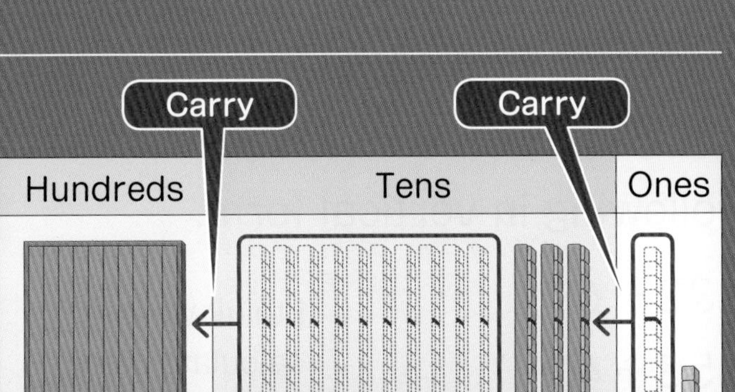

Math Sentence: $74+58=132$

Answer: 132

1 Let's think about how to add $65 + 58$ in vertical form.

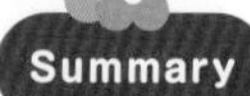

Summary

We can calculate addition with carrying to the tens and the hundreds place in vertical form as the same way as we have learned so far.

2 Let's calculate the following in vertical form.

① $35 + 96$　② $88 + 44$　③ $36 + 89$

④ $54 + 67$　⑤ $51 + 69$　⑥ $32 + 78$

3 Let's think about how to calculate the following in vertical form.

① $37 + 67$　② $6 + 97$　③ $15 + 85$

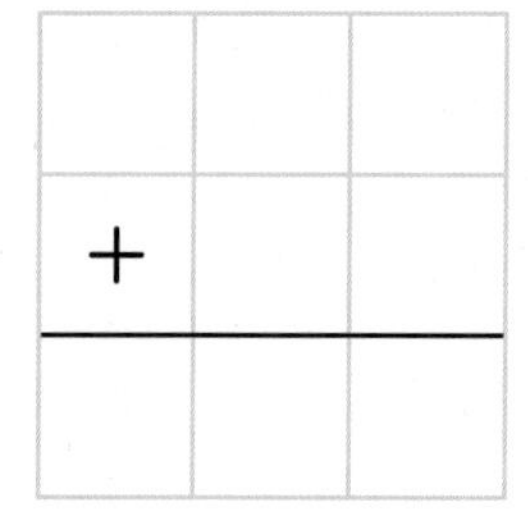

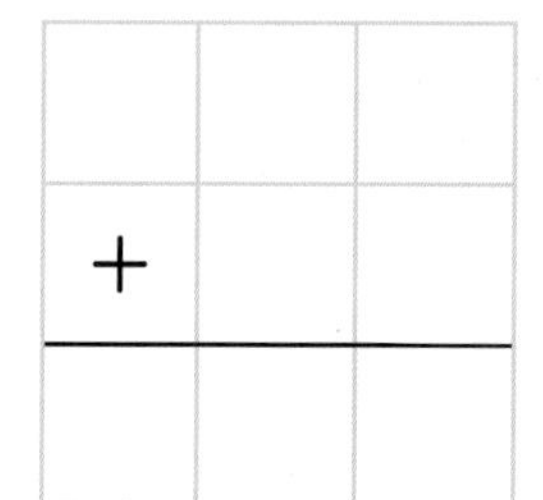

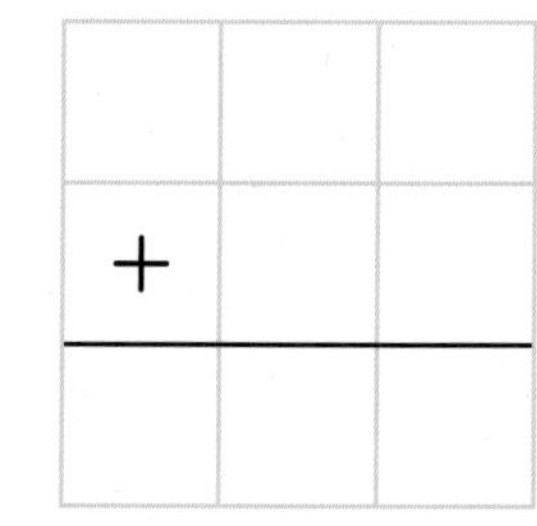

Will 10 tens be carried?

4 Let's calculate the following in vertical form.

① $27 + 78$　② $32 + 69$

③ $51 + 49$　④ $58 + 42$

⑤ $8 + 96$　⑥ $93 + 7$

I think we can calculate using larger numbers.

? Can we do the same for the addition of 3-digit numbers?

2 Addition of 3-digit numbers

1

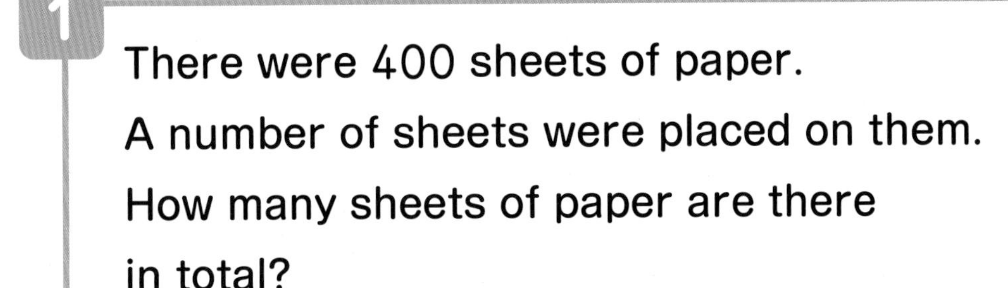

There were 400 sheets of paper.
A number of sheets were placed on them.
How many sheets of paper are there in total?

① 300 sheets of paper were placed. How many sheets are there in total?

Math Expression: ______ Answer: ______ sheets

\ Want to think /

Let's think by using bundles of 100.

Sara

② 300 more sheets of paper are placed on the top of the stack. How many sheets will be there in total?

Math Expression: ______ Answer: ______ sheets

Summary In addition of the numbers that are sets of 100, we can calculate as we have learned so far by using bundles of 100.

Yu

1 Let's calculate the following.

① 100 + 400 ② 200 + 600 ③ 700 + 200

④ 900 + 100 ⑤ 600 + 400 ⑥ 200 + 800

? Can we add 3-digit numbers even when the numbers are not sets of 100?

2 Let's calculate $628 + 7$ in vertical form.

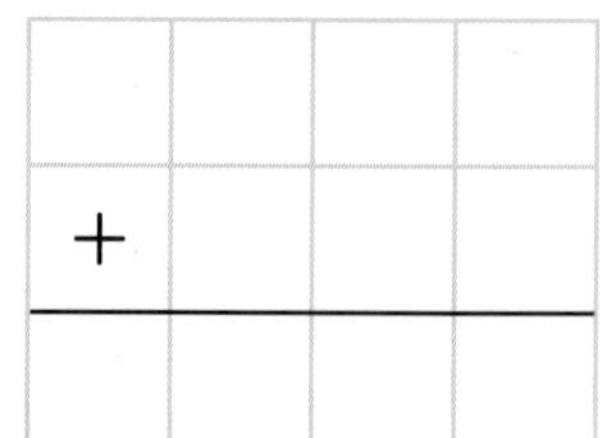

We already learned how to calculate $28 + 7$.

We need to add the numbers in the same places when we calculate in vertical form.

Want to think

Purpose **Can we calculate addition in vertical form for large numbers?**

1 Let's calculate $234 + 57$ in vertical form.

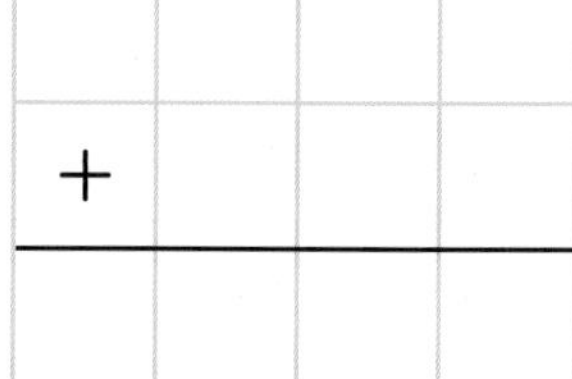

2 Find the mistakes in the following processes. Let's talk about how to calculate correctly.

① $327 + 4$

```
  327
+4
-----
  727
```

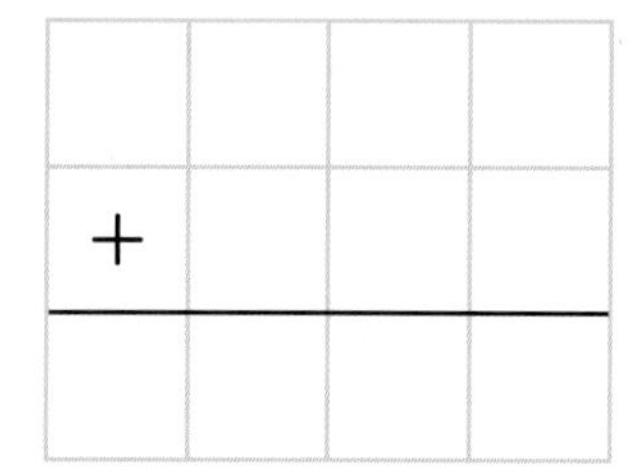

② $649 + 13$

```
  649
+  13
-----
  652
```

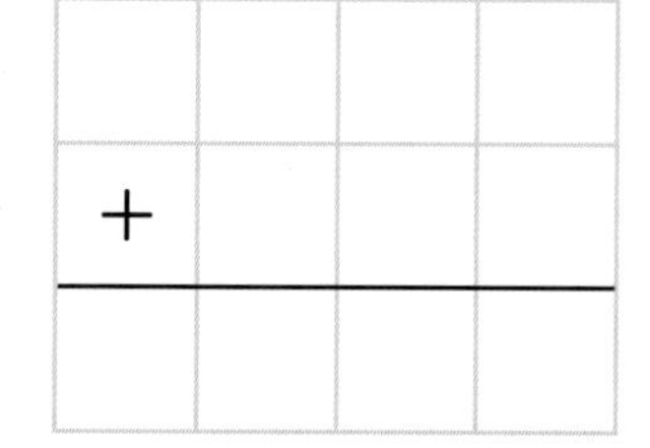

Summary — Way to see and think

Even when the numbers get large, we can add in vertical form by aligning the digits of the numbers according to their places.

3 Let's calculate the following in vertical form.

① $345 + 7$ ② $286 + 4$ ③ $463 + 29$

? Can we also calculate subtraction in large numbers?

3 Subtraction with borrowing from the hundreds place

1 There were 129 colored papers.
I used 73 sheets.
How many sheets are left?

① Let's write a math expression.

About how many sheets will be left?

② Let's think about how to calculate.

Haruto's idea

129 can be decomposed into 100 and 29.
$100-70=30$
$30-3=27$
$29+27=56$

Akari's idea

129 can be decomposed into 120 and 9.
$120-70=50$
$9-3=6$
$50+6=56$

③ Let's explain how to subtract $129-73$ in vertical form.

	1	2	9
−		7	3

On the tens place, we cannot subtract the number written on the lower line from that of the upper line.

Can calculation of large numbers be done as we have learned so far?

Way to see and think

Remember the subtraction of 2-digit numbers in vertical form.

Want to think

Purpose Can we also calculate subtraction of large numbers in vertical form?

Subtraction algorithm for 129 − 73 in vertical form

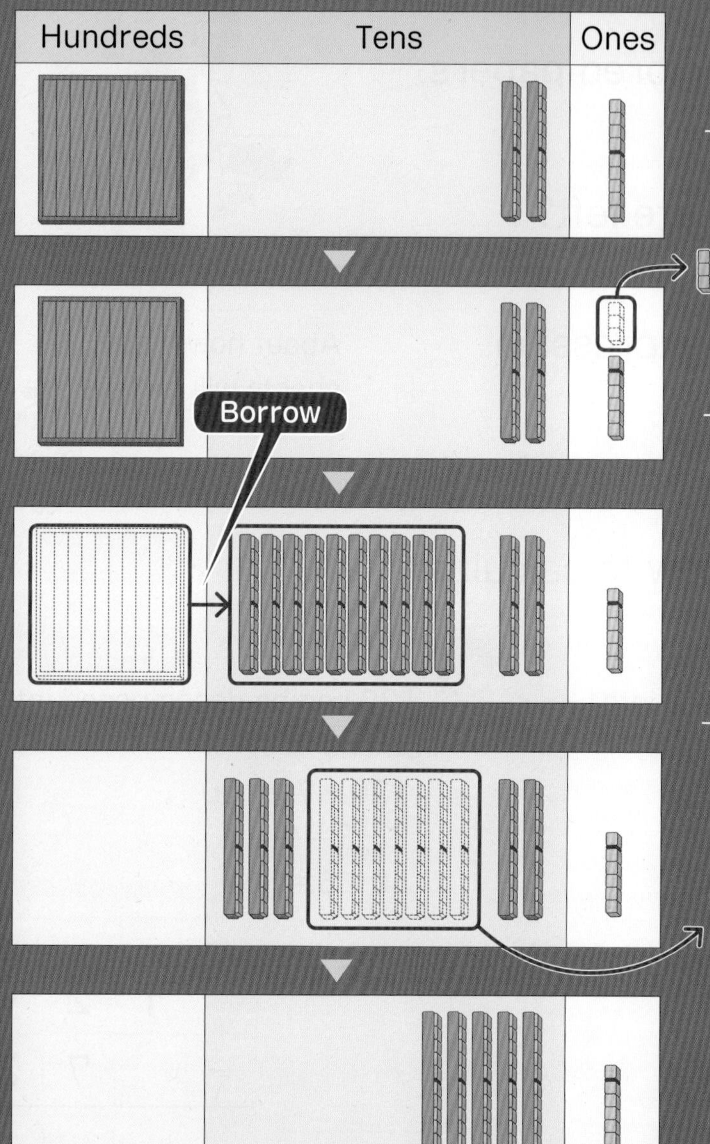

$$\begin{array}{r} 129 \\ -\ 73 \\ \hline \end{array}$$

Align the digits of the numbers according to their places.

$$\begin{array}{r} 129 \\ -\ 73 \\ \hline 6 \end{array}$$

Ones Place

$9-3=6$

The number in the ones place is 6.

$$\begin{array}{r} \overset{10}{\not{1}}29 \\ -\ 73 \\ \hline 56 \end{array}$$

Tens Place

Borrow 1 hundred from the hundreds place as 10 tens, so $12-7=5$.

The number in the tens place is 5.

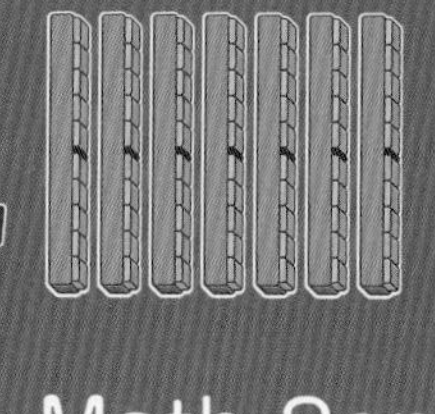

Math Sentence: $129-73=56$

Answer: 56 sheets

The hundreds place turned 0 after borrowing, so we don't need to write the hundreds place.

1 Let's calculate the following in vertical form.

① $135-43$ ② $154-92$

③ $109-53$ ④ $146-60$

? What should I do if both the tens place and the ones place have borrowing?

2

Let's explain how to subtract $125 - 86$ in vertical form.

	1	2	5
−		8	6

Subtraction algorithm for $125 - 86$ in vertical form

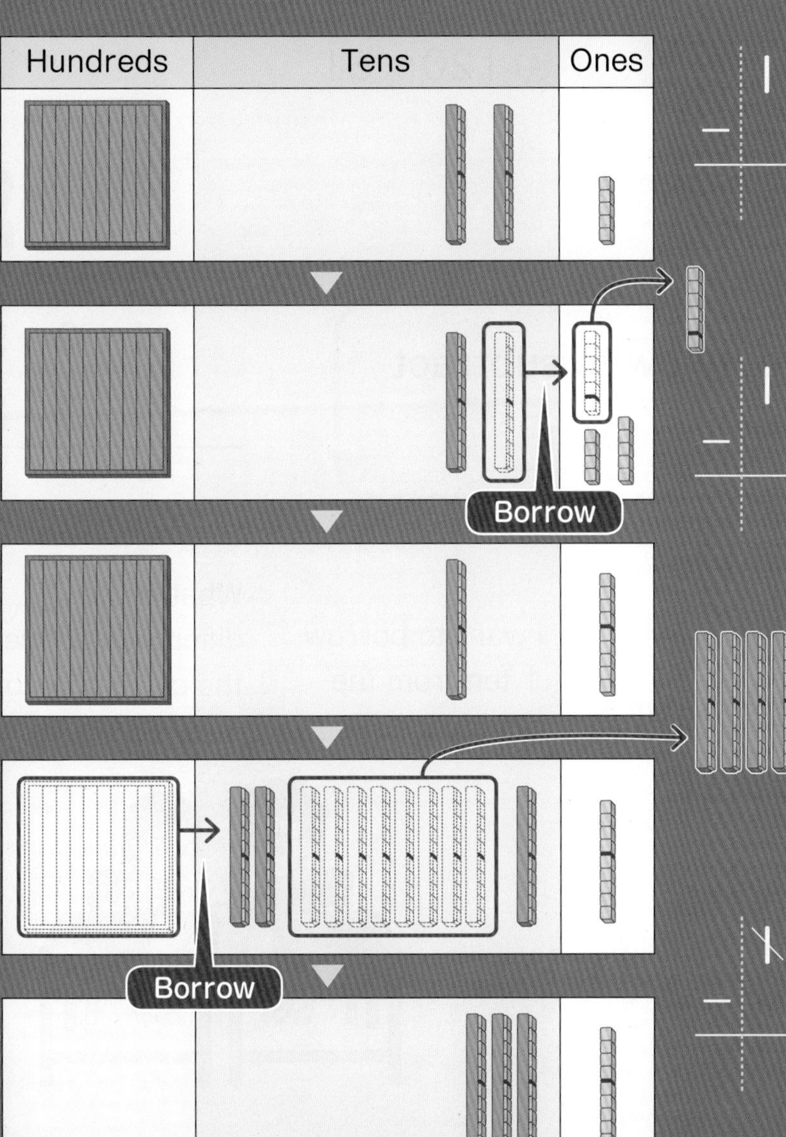

$$\begin{array}{rrr} 1 & 2 & 5 \\ - & 8 & 6 \\ \hline & & \end{array}$$

Align the digits of the numbers according to their places.

$$\begin{array}{rrr} & 1 & 10 \\ 1 & \cancel{2} & 5 \\ - & 8 & 6 \\ \hline & & 9 \end{array}$$

Ones Place

Borrow 1 ten from the tens place as 10 ones, so $15 - 6 = 9$.
The number in the ones place is 9.

$$\begin{array}{rrr} & 10 & \\ & 1 & 10 \\ \cancel{1} & \cancel{2} & 5 \\ - & 8 & 6 \\ \hline & 3 & 9 \end{array}$$

Tens Place

Borrow 1 hundred from the hundreds place as 10 tens, so $11 - 8 = 3$.
The number in the tens place is 3.

Math Sentence: $125 - 86 = 39$　Answer: 39

Summary

Even when the numbers get large, we can subtract in vertical form by aligning the the digits of the numbers according to their places.

1 Let's calculate the following in vertical form.

① 156 − 78 ② 171 − 82

③ 145 − 59 ④ 120 − 61

? Can we borrow under any situation?

3

Let's think about how to subtract 105 − 78 in vertical form.

	1	0	5
−		7	8

① Let's explain how to subtract by using math expressions or diagrams.

Hundreds	Tens	Ones

I want to borrow 1 ten from the tens place...
What is the difference between the calculation of 125 − 86?

\ Want to know /

When 1 ten cannot be borrowed from the tens place, how can we subtract in vertical form?

Subtraction algorithm for $105-78$ in vertical form

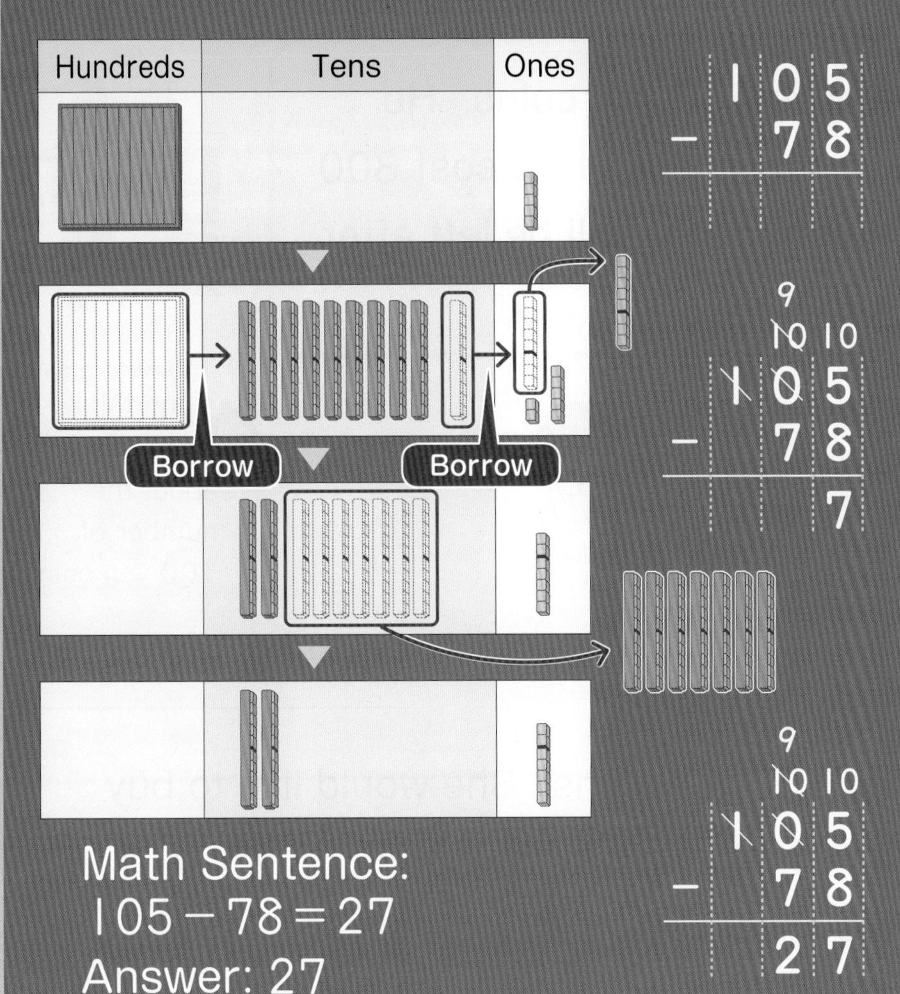

Align the digits of the numbers according to their places.

Ones Place

First borrow 1 hundred from the hundreds place as 10 tens, and then borrow 1 ten from the tens place as 10 ones. So $15-8=7$. The number in the ones place is 7.

Tens place

$9-7=2$

The number in the tens place is 2.

Math Sentence:
$105-78=27$
Answer: 27

Summary

When 1 ten cannot be borrowed from the tens place, borrow 1 hundred from the hundreds place to the tens place as 10 tens, and then borrow 1 ten from the tens place to the ones place as 10 ones.

1 Let's calculate the following in vertical form.

① $106-59$ ② $100-36$ ③ $102-7$

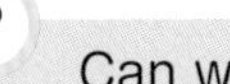

? Can we subtract larger 3-digit numbers?

4 Subtraction of 3-digit numbers

1

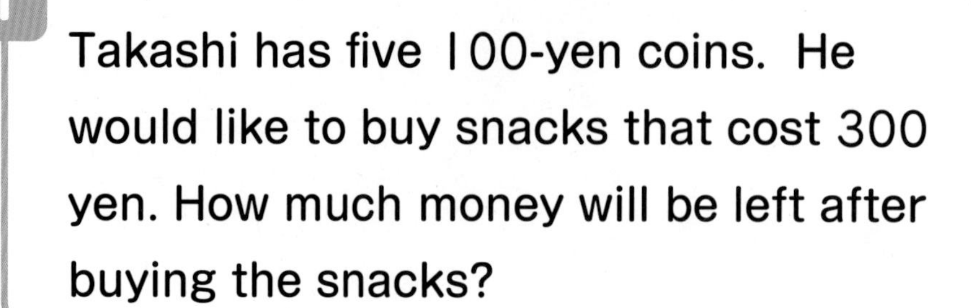

Takashi has five 100-yen coins. He would like to buy snacks that cost 300 yen. How much money will be left after buying the snacks?

\ Want to think /

Purpose As we did in addition, let's think as 100 as one set.

Akari

Way to see and think

Let's think about the number of 100-yen coins.

Math Sentence: ☐ Answer: ☐ yen

1 Marin has ten 100-yen coins. She would like to buy snacks that cost 300 yen. How much money will be left after buying the snacks?

Math Sentence: ☐ Answer: ☐ yen

Summary If we set 100 as 1 unit, we can calculate subtractions of 3-digit numbers as we learned.

Haruto

2 Let's calculate the following.

① $900 - 500$ ② $500 - 200$ ③ $600 - 400$

④ $700 - 100$ ⑤ $800 - 300$ ⑥ $1000 - 200$

? Can we do other subtractions with 3-digit numbers?

2 Let's calculate 753 − 6 in vertical form.

We have already learned how to calculate 53 − 6.

We need to align the digits...

Want to think

Purpose **Can we calculate subtraction of large numbers in vertical form?**

1 Let's calculate 546 − 27 in vertical form.

2 Find the mistakes in the following processes. Let's talk about how to calculate correctly.

① 608 − 3

```
 608
-3
----
 308
```

② 524 − 17

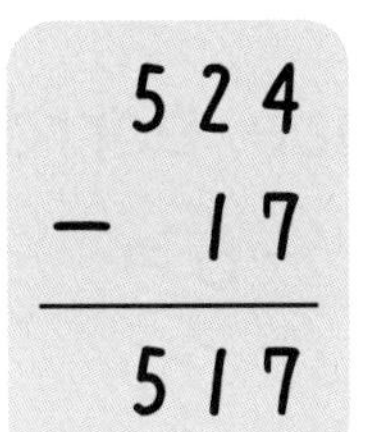

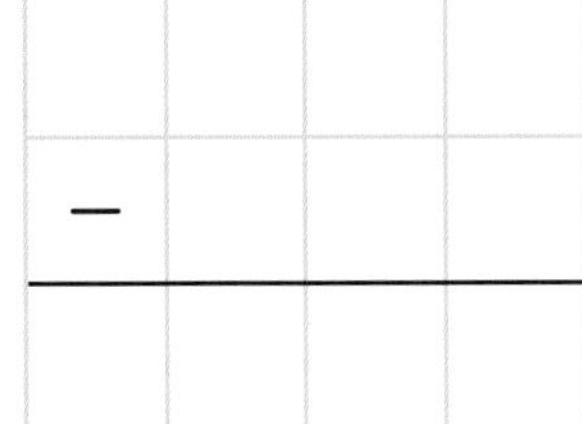

Way to see and think

Even when the numbers get large, we can subtract in vertical form as we have learned by aligning the digits of the numbers according to their places.

3 Let's calculate the following in vertical form.

① 536 − 5　② 273 − 4　③ 115 − 8

④ 354 − 32　⑤ 282 − 63　⑥ 230 − 24

CAN What can you do?

☐ We understand the meaning of calculation and the way to calculate in vertical form. → pp.101 ~ 102, pp.109 ~ 110

1 Let's summarize how to calculate the following in vertical form.

① $73+56$

(1) In the ones place, $3+6=9$

(2) In the tens place, $7+5=12$

The tens place is ☐. Carry ☐ hundred to the hundreds place.

(3) The answer is ☐.

+		

② $132-64$

(1) Borrow ☐ ten from the tens place, so the ones place will become ☐ $-4=$ ☐.

(2) In the tens place, borrow ☐ hundred from the hundreds place, so ☐ $-6=$ ☐.

(3) The answer is ☐.

−			

☐ Can calculate in vertical form. → pp.101 ~ 113

2 Let's calculate the following in vertical form.

① $73+45$ ② $69+71$ ③ $46+55$ ④ $374+6$

⑤ $148-67$ ⑥ $168-79$ ⑦ $107-48$ ⑧ $114-9$

Supplementary Problems → p.157

Which "Way to See and Think Monsters" did you find in "9 Addition and Subtraction of Large Numbers"?

I found "Same Way" when I was trying subtraction.

I found other monsters too!

Usefulness and Efficiency of Learning

1 Find the mistakes in the following processes. Let's write the correct answer in the ( ).

①
$$\begin{array}{r} 98 \\ +\ \ 9 \\ \hline 117 \end{array}$$
(　　　　)

②
$$\begin{array}{r} 121 \\ -\ 68 \\ \hline 63 \end{array}$$
(　　　　)

③
$$\begin{array}{r} 105 \\ -\ 17 \\ \hline 98 \end{array}$$
(　　　　)

④
$$\begin{array}{r} 168 \\ -\ 97 \\ \hline 131 \end{array}$$
(　　　　)

2 Yuri has 73 sheets of origami paper, and her sister has 89 sheets.

① How many sheets of origami paper do they have in total?

② Who has more sheets and by how many?

3 Let's summarize how to calculate 200 + 700.

(1) 200 is a number that is the sum of ☐ sets of 100.

700 is a number that is the sum of ☐ sets of 100.

(2) There are ☐ sets of 100 in total.

(3) The answer is ☐.

4 Let's make various addition expressions to make 1000 by using two numbers that are sets of 100.

With the Way to see and think Monsters...

Let's Reflect!

Let's reflect on which monster you found in " 9 Addition and Subtraction of Large Numbers."

Align

Even when the numbers get large, we could calculate addition and subtraction by aligning the digits of the numbers according to their places.

Same Way

Even when the numbers get large, we could do carrying and borrowing as we learned.

① Compare addition and subtraction of 2-digit and 3-digit numbers in vertical form to addition and subtraction in vertical form which you have learned. Talk to your friends about what you found out.

2-digit + 2-digit

$$\begin{array}{r} 38 \\ +\ 27 \\ \hline {}^{1}\ \ \\ 65 \end{array}$$

2-digit − 2-digit

$$\begin{array}{r} {}^{3}\ {}^{10} \\ \not{4}5 \\ -\ 27 \\ \hline 18 \end{array}$$

2-digit + 2-digit

$$\begin{array}{r} 74 \\ +\ 58 \\ \hline {}^{1}\ \ \\ 132 \end{array}$$

3-digit − 2-digit

$$\begin{array}{r} {}^{9}\ \ \\ {}^{\not{10}}\ {}^{10} \\ \not{1}\not{0}5 \\ -\ \ 78 \\ \hline 27 \end{array}$$

In any calculation using vertical form, we align the [ ].

Sara

Let's deepen. → p.160

? Solve the ?

We could calculate addition and subtraction of large numbers in vertical form as we learned.

Haruto

→

Want to Connect

Can we calculate (3-digit) + (3 digit) and (3-digit) − (3-digit) in the same way?

Yu

To what should we pay attention when we subtract in vertical form?

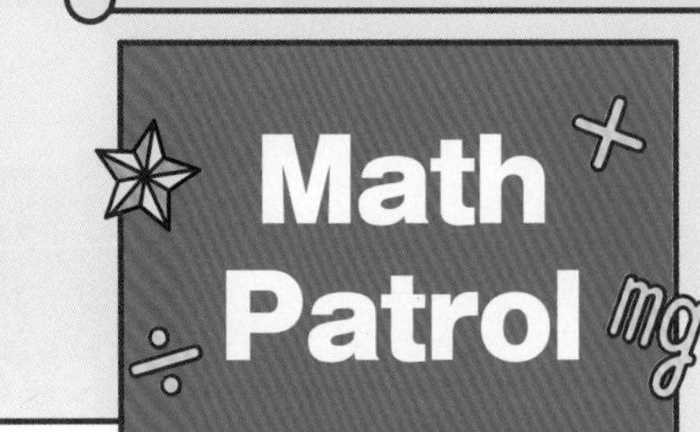


```
  3 2 2
−   1 4
  3 1 2
```

I made a mistake in the ones place. I did 4 − 2.

```
    1 10
  3 2̸ 2
−   1 4
  3 0 8
```

We can't calculate 2 − 4 !

Subtracting the number written on the lower line from that on the upper line.

Frequently made mistake ①
Not subtracting the number written on the lower line from that on the upper line, but subtracting the smaller number from the larger number.

Be careful!
When you can't subtract the number written in the lower line from that on the upper line, borrow from the higher place.

```
    10 10
  1̸ 0̸ 5
−    6 8
     4 7
```

I forgot that I borrowed one 10 from the tens place to the ones place.

```
     9
    1̸0 10
  1̸ 0̸ 5
−    6 8
     3 7
```

Write down that 10 became 9 by borrowing!

Frequently made mistake ②
Subtracting the number written on the lower line from 10, forgetting that you have borrowed 10 from the tens place to the ones place.

Be careful!
Let's draw a diagonal line or write down the number so that you won't forget that you have done the borrowing.

Even when you make a mistake, you can always confirm your answer.

```
    3 7
+   6 8
  1 0 5
```

Reflect

Connect

Problem Let's compare addition and subtraction in vertical form.

Let's calculate 59 + 73 in vertical form.

$$\begin{array}{r} 59 \\ +\ 73 \\ \hline 132 \end{array}$$

① Align the digits of the numbers according to their places.

② Ones Place

9 + 3 = ① 2 — as 10 ones

Carry (1 ten) to the tens place.

③ Tens Place

5 + 7 + 1 = ① 3 — as 10 tens

Carry (1 hundred) to the hundreds place.

In fact it means
50 + 70 + 10 = ① 30

Hundreds Place

Add the numbers in the same places.

addition of 1-digit numbers

5 + 7 + 1 = 13
The answer is 13.
Why is 1 hundred carried to the hundreds place?

In "5 + 7 + 1," each number is in the tens place. So it means "50 + 70 + 10."

Let's subtract 146 − 78 in vertical form.

```
   10
    3 10
  1̸ 4̸ 6
−   7 8
―――――――
    6 8
```

① Align the digits of the numbers according to their places.

② Ones Place

Borrow (1 ten) to the tens place. — as 10 ones

$16 - 8 = 8$

③ Tens Place

Borrow (1 hundred) to the hundreds place. — as 10 tens

$13 - 7 = 6$

In fact it means $130 - 70 = 60$.

Summary

Same Way

· To calculate the numbers in the same places.

· To calculate in ones place first.

Difference

· To carry in addition, to borrow in subtraction.

Let's look back

○ In vertical form, the calculation in each place is the same as the one we have learned in the 1st grade.

○ It seems that calculation of large numbers can be done as we have learned.

Calculation in the same places is addition or subtraction of 1-digit numbers that we have learned in the 1st grade.

By aligning the digits of the numbers according to their places, we can calculate the numbers in the same places.

I think we can find the answers of the calculation of much larger numbers by calculating in the same places.

Utilizing Math for SDGs

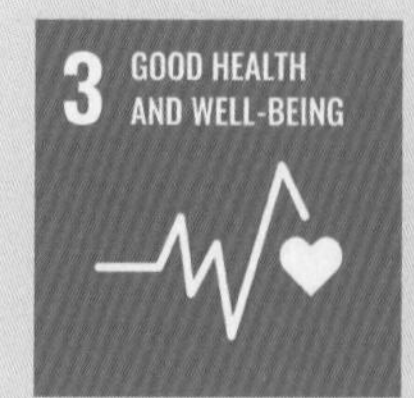

Summer is coming!
How can we spend a happy and healthy summer vacation?

Summer vacation is coming soon.
In summer, you may feel tired or sluggish because of the heat.
Sometimes you may suffer from chronic fatigue.
In order to avoid these, solutions such as the following might be effective.

- take good rest at night
- take balanced dietary meals
- continue moderate exercise

To make your summer vacation more enjoyable, let's think about effective rules.

Nishioyama Station,
Kagoshima Prefecture

Cape Hirakubo,
Ishigaki Island

Sara

I heard that chronic fatigue is really dangerous.
To avoid being exposed to it, we wear caps and put something to drink in our water bottles when we go outside to play.

① Write down three rules you would do to make your summer vacation a happy and healthy one.

Haruto

Since I tend to play games too long, I will set the time for the playing games.

Yu

Walking my dog can be an exercise that I can continue.

Akari

I will get up at the same time as when I had school.

② Let's make a presentation on what you have written in ①.

③ Let's make a daily plan from 6 a.m. to 9 p.m. to make your summer vacation a healthy one.

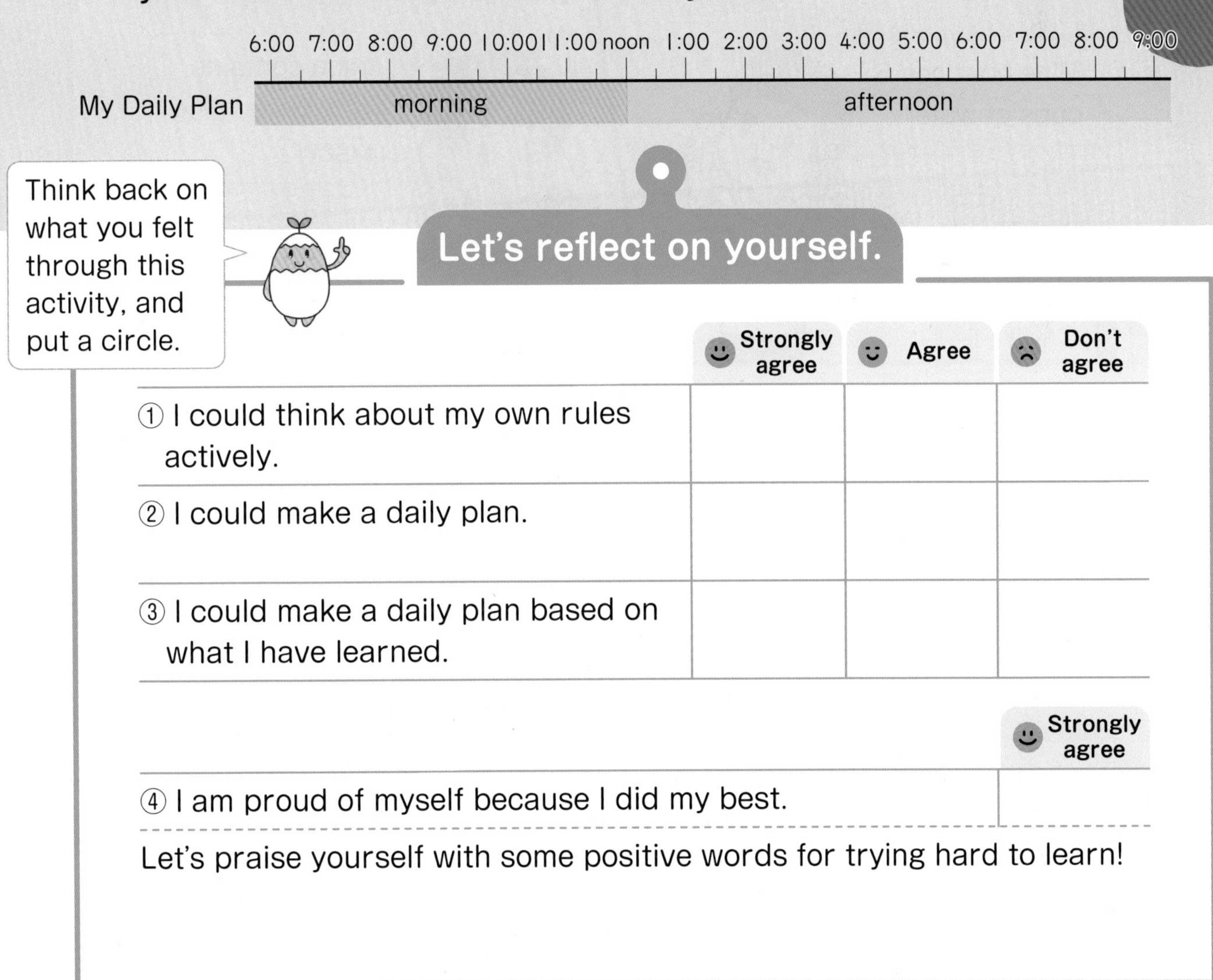

	6:00	7:00	8:00	9:00	10:00	11:00	noon	1:00	2:00	3:00	4:00	5:00	6:00	7:00	8:00	9:00
My Daily Plan	morning						afternoon									

Think back on what you felt through this activity, and put a circle.

Let's reflect on yourself.

	Strongly agree	Agree	Don't agree
① I could think about my own rules actively.			
② I could make a daily plan.			
③ I could make a daily plan based on what I have learned.			

	Strongly agree
④ I am proud of myself because I did my best.	

Let's praise yourself with some positive words for trying hard to learn!

Find the ?

Which is more?

\ Want to compare /

Purpose Let's compare which bottle can hold more water.

10 Amount of Water

Let's think about how to compare the amounts of water and how to represent them.

1 How to compare the amounts of water

How to compare the amounts of water ↓

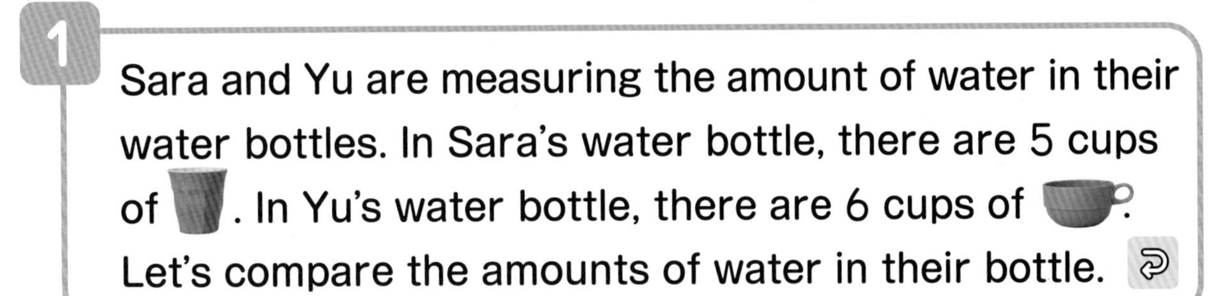

1 Sara and Yu are measuring the amount of water in their water bottles. In Sara's water bottle, there are 5 cups of [cup]. In Yu's water bottle, there are 6 cups of [cup]. Let's compare the amounts of water in their bottle.

① Can we say which one holds more?

Akari
Is 6 cups more than 5 cups?

But the size of the cups...
Haruto

A quantity of something like water is called an "amount."

② Let's try by using the same cup. Which one contains more?

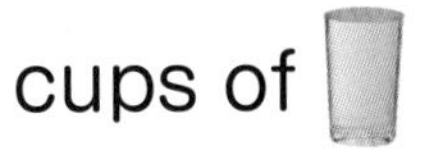

☐ cups of [cup]

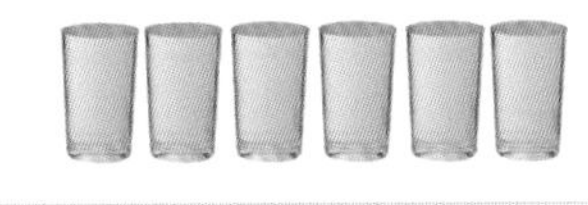

☐ cups of [cup]

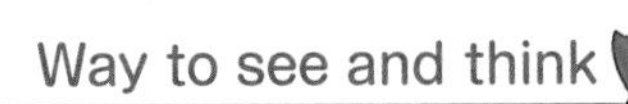

Summary

We can compare amounts by the number of units of amount used.

? What can we do to compare the more amounts of water without using cups?

2 How to represent the amount of water

1 Let's measure the amount of water in the kettle.

Sara
How about using a lot of cups?

? Purpose ＼ Want to know ／ **Do we also have units for measuring amounts?**

We can compare the amount of water by counting the number of filled **unit cups**.

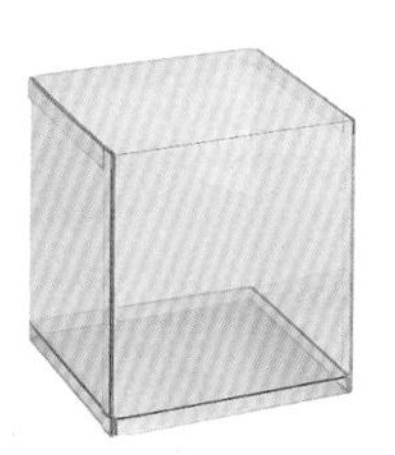

There is a unit called **liter** to represent the amount of water.

1 liter is written as 1 L.

Each container in the picture above can hold 1 L.

In many countries, liter is used as a unit to measure the amount of liquid.

Way to see and think
We can represent the amount of water by measuring how many 1 L there are.

① The amount of water in the kettle was measured by using 1 L measuring cups. It contained three.
How many liters is the amount of water in the kettle?

1 The following amount of water was measured by using 1 L measuring cups. How many liters is each amount of water?

① Plastic bottle

☐ L

② Bucket

☐ L

Summary

We can measure and compare the amount of liquid by using L as the unit of measure.

2 Let's look for containers with L as the unit of measure.

? How can we measure the portion that is less than 1 L?

Portion that is less than 1L→

2

The amount of water in the pot was measured by using 1 L measuring cups. How can we represent the portion that is less than 1 L?

When we learned length, we divided 1 cm into 10 lengths as mm.

Akari

Want to know

? **Purpose** **Is there a unit that represents the amount of water that is less than 1 L?**

To measure the portion that is less than 1 L, we can use a 1 **deciliter** measuring cup.

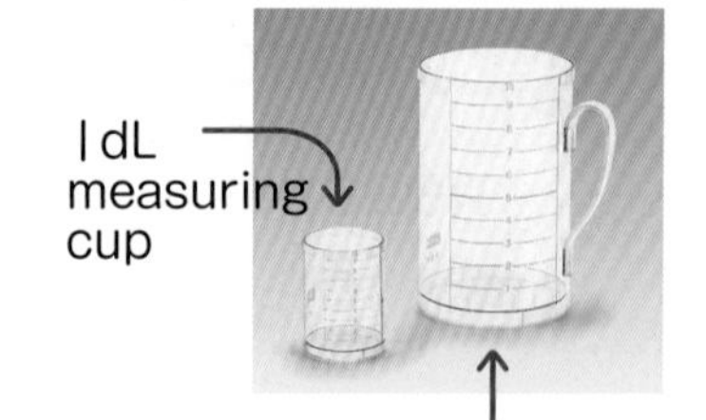

1 L measuring cup

1 Let's fill in a 1 L measuring cup with water by using a 1 deciliter measuring cup. How many times do we need to use the cup?

When 1 L is divided into 10 equal amounts, the unit of 1 amount is called deciliter. 1 deciliter is written as 1 dL. Deciliter is another unit to represent the amount of liquid.

▷ 1L = 10dL

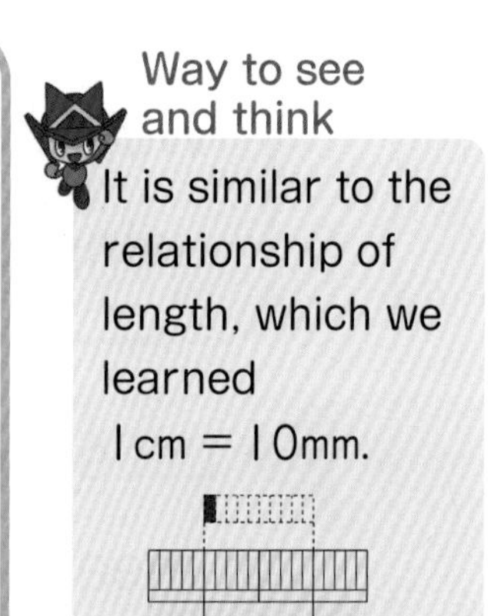

Way to see and think
It is similar to the relationship of length, which we learned
1 cm = 10mm.
1 cm

2 The amount of water in the pot in **2** was 2 L and three 1 dL measuring cups. How can we represent the amount of water using L and dL?

Summary

The amount that is less than 1 L can be represented by using the unit "dL".

3 How many dL are contained in the container?

4 Let's measure the amount of water that can be placed into the following containers.

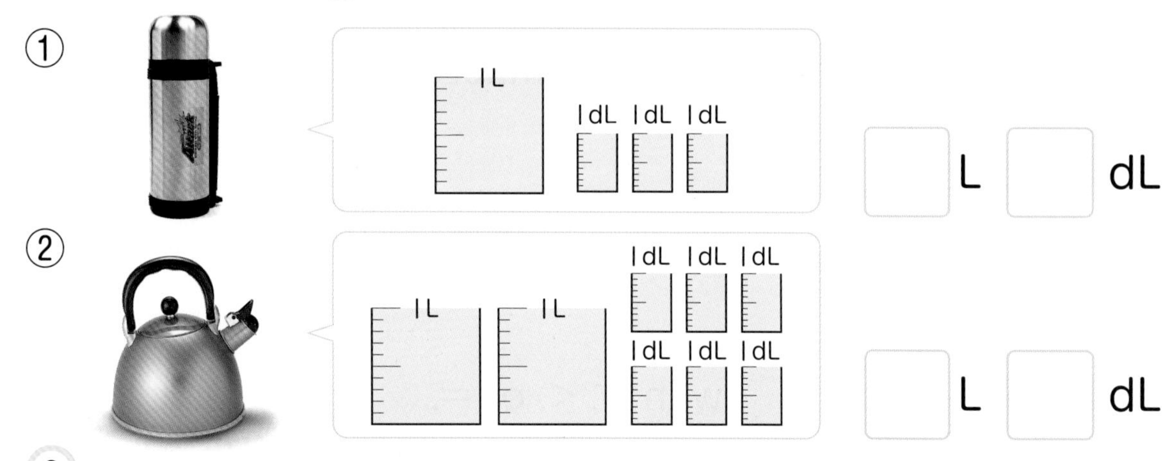

? Can we represent the amount of water in two different ways like we could in length?

3 Let's find out the amount of water in a bigger container like the one on the right.

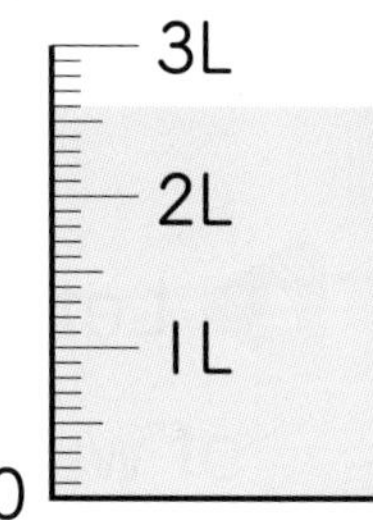

When we learned length, we could represent the same length in two ways such as 3cm8mm and 38mm.

Want to represent

Purpose Can we represent the amount of water in two ways?

❶ How many L and dL are there?

2L and 6 smaller scales is

L	dL
2	6

❷ How many dL are there?

2L = ☐ dL, so when it is added to 6 dL, the total is ☐ dL.

	dL
2	6

1 How many L and dL are there in 54dL?

Summary We can represent amount of water in two ways.

2 Let's fill in the ☐ with numbers.

① 3 L = ☐ dL ② 4 L 9 dL = ☐ dL

③ 68 dL = ☐ L ☐ dL ④ 2 L 1 dL = ☐ dL

3 Let's fill in the ☐ with >, <, or =.

① 2 L 5 dL ☐ 1 L 5 dL ② 1 L 3 dL ☐ 13 dL

③ 70 dL ☐ 7 L 2 dL

4 Let's estimate the amount of water in various containers and measure them.

Container	Estimated amount	Actual amount
Water bottle	5dL	5dL
Pan	1 L	1 L 2dL

? How can we represent the amount of water that is less than 1 dL?

That's it! Making a 1 dL Measuring Cup ▷

Fill the 1 dL cup with water, pour the water into a container, and put a mark on the container to indicate where the water level is.

4 Let's use a 1 dL measuring cup to find out the amount of juice in the can.

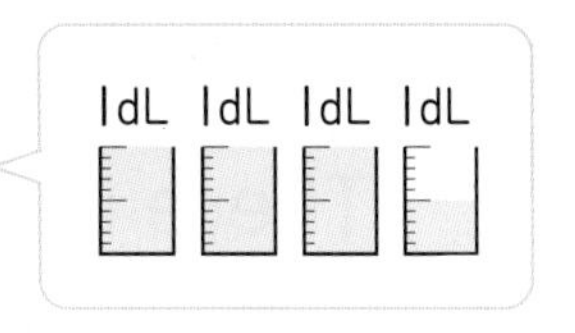

Want to know

Purpose Is there a unit that represents the amount of water that is less than 1 dL?

To represent the amount of water, there is a unit called **milliliter** that is smaller than L and dL.

1 milliliter is written as 1 mL.

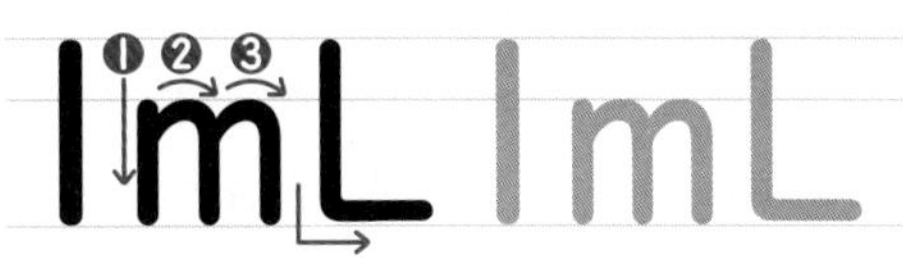

1 Let's use a 1 L measuring cup and 1 dL measuring cups to find out the amount of water in a 1000 mL pack.

1 L=1000mL	1 dL=100mL

We can write liter as L, l, or ℓ. Milliliter can also be written as mL, ml, or mℓ. 1 mL can also be called 1 cc.

Summary The amount that is less than 1 dL can be represented with mL.

2 Let's look for containers with mL as the unit of measure.

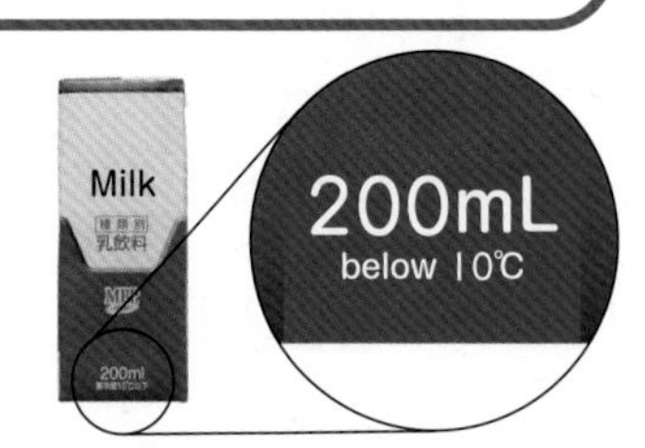

Can we calculate the amount like we did in length?

3 Calculating the amount of water

1 There is 1L 3 dL of water in a plastic bottle and 2L 5dL of water in a bottle.

① How many L and dL are there in total?

Math Expression: ______

Haruto
Can we calculate like we could in length?

Want to think

Purpose How can we calculate the amount of water?

② Let's think about how to calculate.

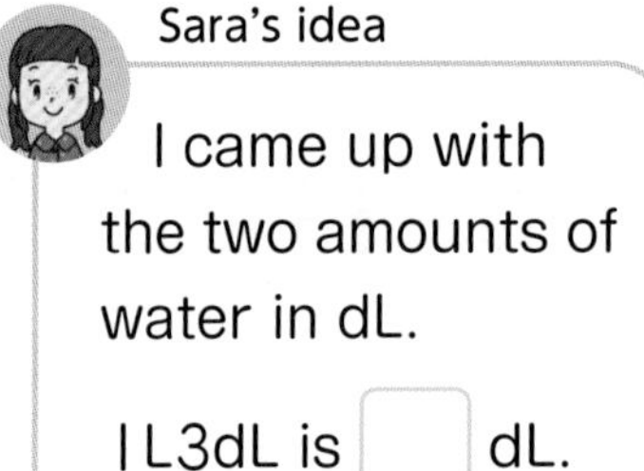

Sara's idea
I came up with the two amounts of water in dL.
1L3dL is ☐ dL.
2L5dL is ☐ dL.

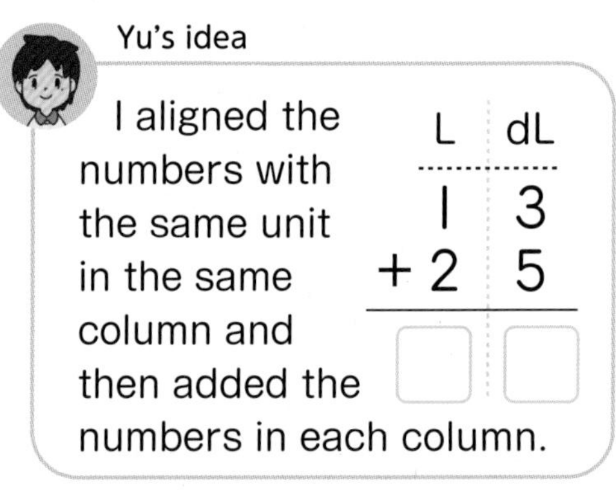

Yu's idea
I aligned the numbers with the same unit in the same column and then added the numbers in each column.
L dL
1 3
+ 2 5
☐ ☐

③ What is the difference between the amount of water in the plastic bottle and the bottle?

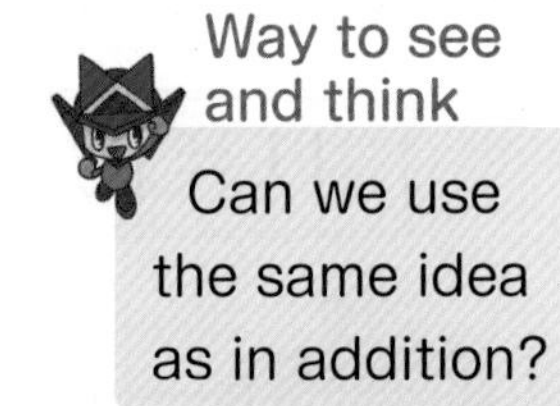

Way to see and think
Can we use the same idea as in addition?

Summary We can calculate the amount of water by aligning the numbers with the same unit.

1 Let's calculate the following.

① 3L2dL + 1L5dL ② 3L6dL + 1L8dL

③ 6L4dL − 1L3dL ④ 7L − 3L5dL

CAN What can you do?

☐ We can represent the amount of water. → pp.127 ~ 128

1 Which measuring cup would be appropriate to measure the amount of water of the following containers, 1 L cups or 1 dL cups?

① rice bowl

☐ measuring cups

② basin

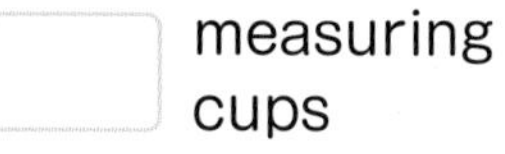

☐ measuring cups

☐ We understand the conversion of the units of amount. → p.126, p.129

2 Let's fill in the ☐ with numbers.

① 1 L = ☐ dL ② 1 L = ☐ mL ③ 1 dL = ☐ mL

☐ We can calculate the amount of water. → p.130

3 There are some water in two containers.

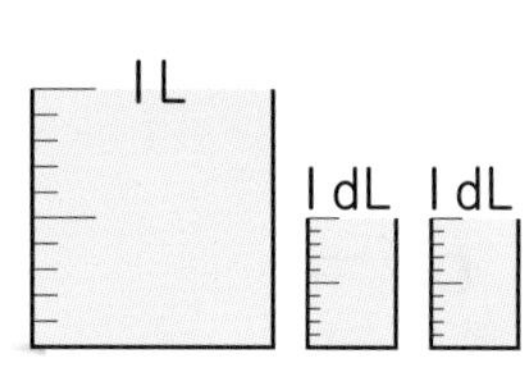

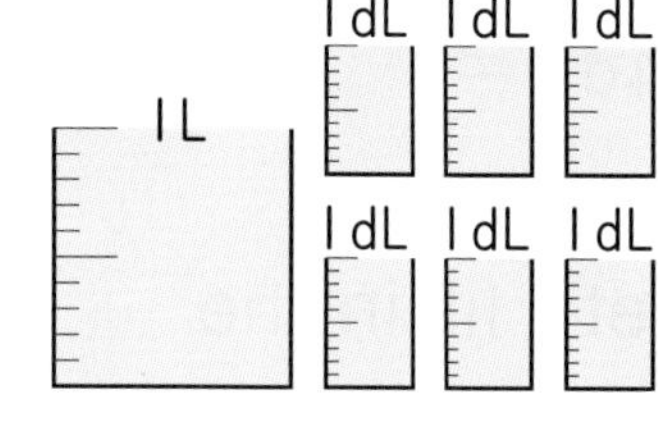

① How many L and dL are there in total?

② What is the difference between the amount of water in each container?

Supplementary Problems → p.158

Which "Way to See and Think Monsters" did you find in "10 Amount of Water"?

Yu: I found "Unit" when I compared the amount of water.

Akari: I found other monsters too!

Utilize Usefulness and Efficiency of Learning

1 What is the amount of water?

① 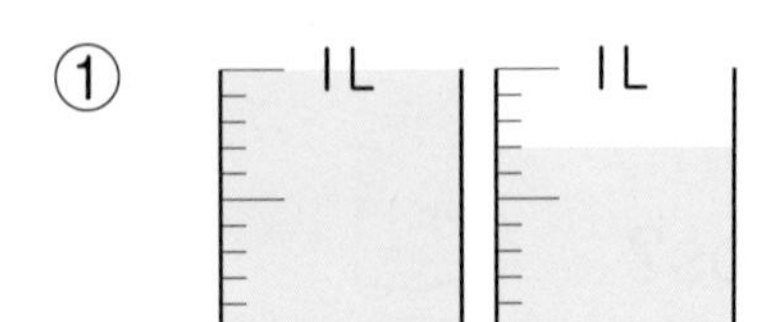

② 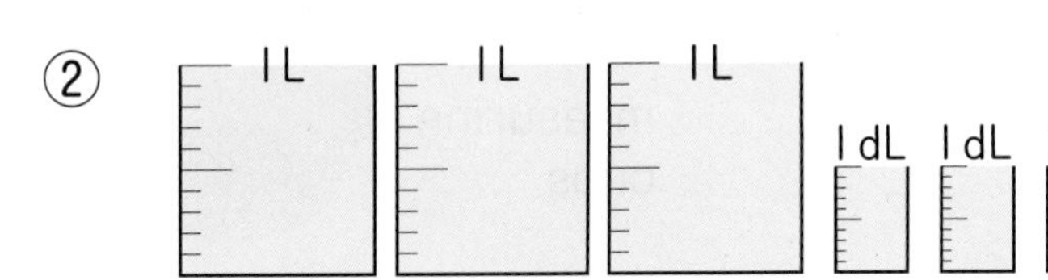

③ 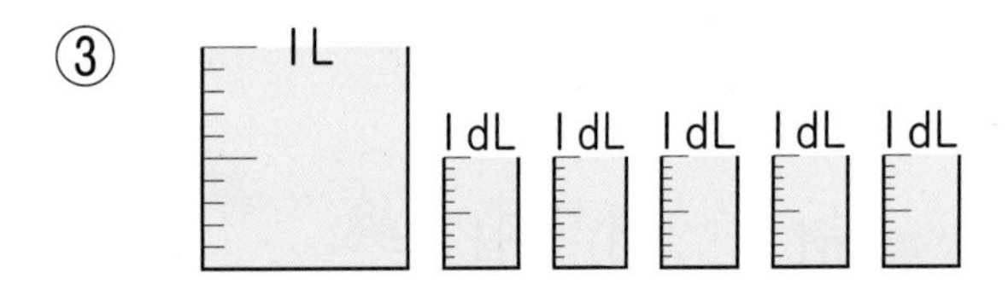

☐ L ☐ dL

2 Let's fill in the ☐ with numbers.

① 5L= ☐ dL　　② 75dL= ☐ L ☐ dL

③ 4dL= ☐ mL　　④ 3L8dL= ☐ dL

3 Let's fill in the ☐ with >, <, or =.

① 9dL ☐ 8dL　　② 2L ☐ 20dL

③ 450mL ☐ 5dL　　④ 3L2dL ☐ 31dL

4 Let's calculate the following.

① 2L+1L5dL　　② 1L3dL+7dL

③ 4L5dL+1L6dL　　④ 3dL+2L8dL

⑤ 3L6dL−1L5dL　　⑥ 2L4dL−4dL

⑦ 6L4dL−2L6dL　　⑧ 1L−7dL

With the Way to see and think Monsters...

Let's Reflect!

Let's reflect on which monster you found in "10 Amount of Water."

Unit

We could represent the amount of water by setting 1 L, 1 dL, 1 mL as one unit.

① What is the difference between representing the amount of water as "how many cups" and "how many liters?"

We can represent by using how many cups, but the size of the cup could be different. Also we might not have enough cups.

Haruto

By using L, we can tell the amount of water with the same unit.

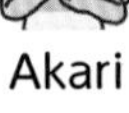
Akari

Same Way

In the same way as we learned in the unit of length where ten 1 mm makes 1 cm, ten 1 dL makes 1 L in representing the amount of water.

② What is the similarity in representing the length and the amount of water?

In length, we represented 1 cm by dividing 1 cm into 10 parts of 1 mm.

Sara

In amount of water, we could represent 1 L by dividing 1 L into ten 1 dL too.

Yu

Let's deepen. → p.161

Solve the ?

By using units such as L, we can compare amount of water and represent it in numbers anytime.

Sara

Want to Connect

Are there units other than amount of water and length?

Akari

11 Triangles and Quadrilaterals

Let's examine the shapes and sort them.

1 Triangles and quadrilaterals

Let's connect the points by using straight lines to enclose the animals.

1 Let's sort the shapes that are enclosed by straight lines into 2 groups.

How can we categorize them?

What is the number of straight lines?

Way to see and think

What is the similarity and difference between the two groups?

? Purpose What kinds of shapes can you find?

I categorized based on the number of straight lines that are used.
Yu
Way to see and think
You can sort by the number of straight lines.

The shape enclosed by 3 straight lines is called a **triangle**.
The shape enclosed by 4 straight lines is called a **quadrilateral** (or **quadrangle**).

? What are the properties of a triangle and a quadrilateral?

2 Let's find triangles and quadrilaterals.

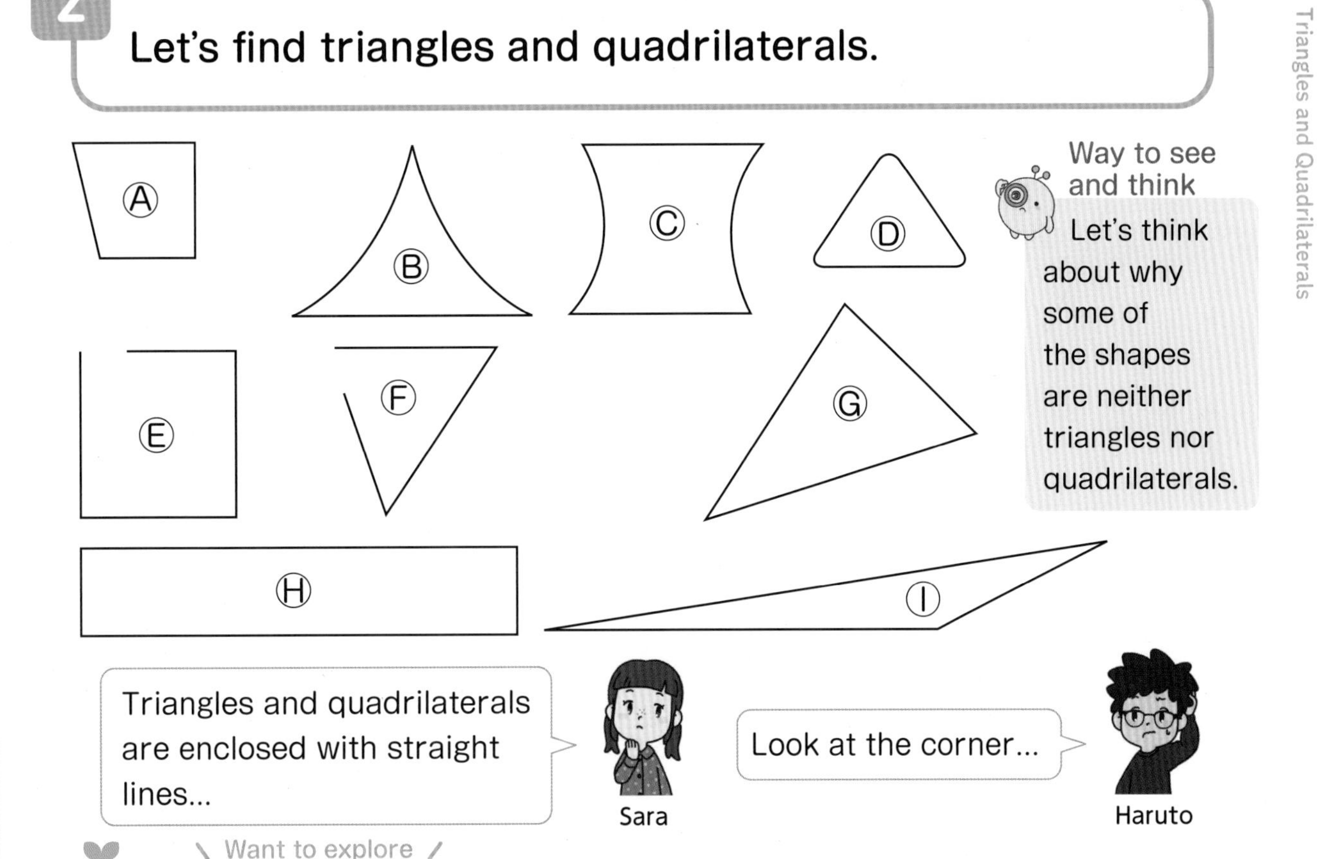

Way to see and think
Let's think about why some of the shapes are neither triangles nor quadrilaterals.

Sara: Triangles and quadrilaterals are enclosed with straight lines...

Haruto: Look at the corner...

Want to explore

Purpose What are the properties of a triangle and a quadrilateral?

Each straight line around a triangle and a quadrilateral is called a **side** and each corner point that two sides make is called a **vertex**.

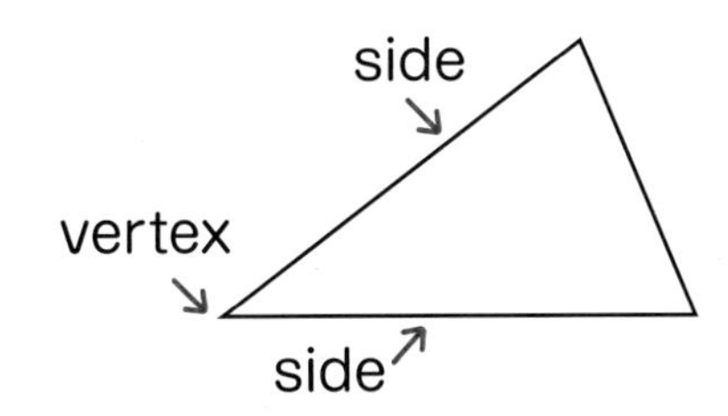

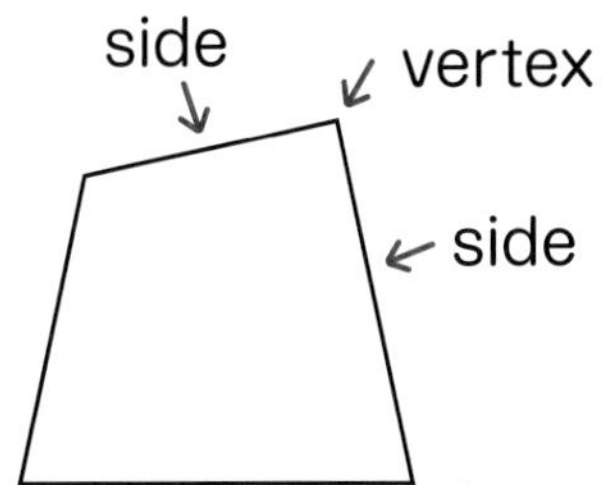

In a triangle, the number of sides is ☐ and the number of vertices is ☐.

In a quadrilateral, the number of sides is ☐ and the number of vertices is ☐.

1 Let' s draw various triangles and quadrilaterals by connecting points with straight lines.

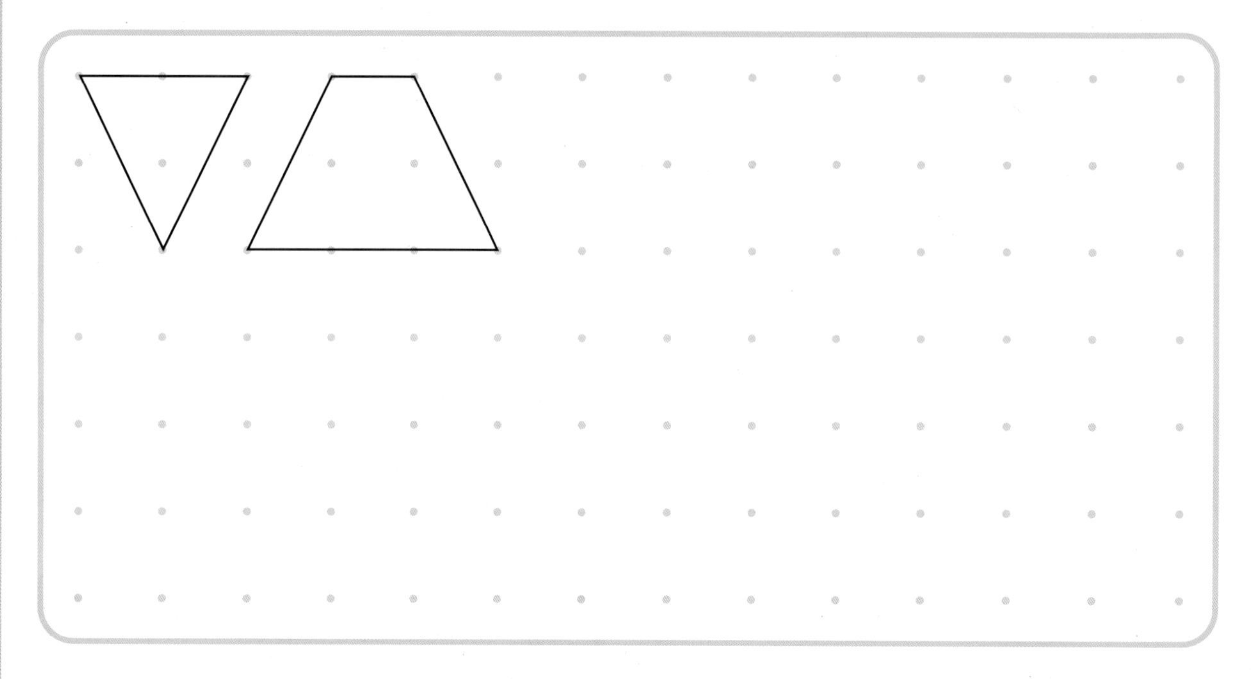

Summary Triangle is a shape that has 3 sides and 3 vertices. Quadrilateral is a shape that has 4 sides and 4 vertices.

Haruto

2 Let's find triangles and quadrilaterals and trace them.

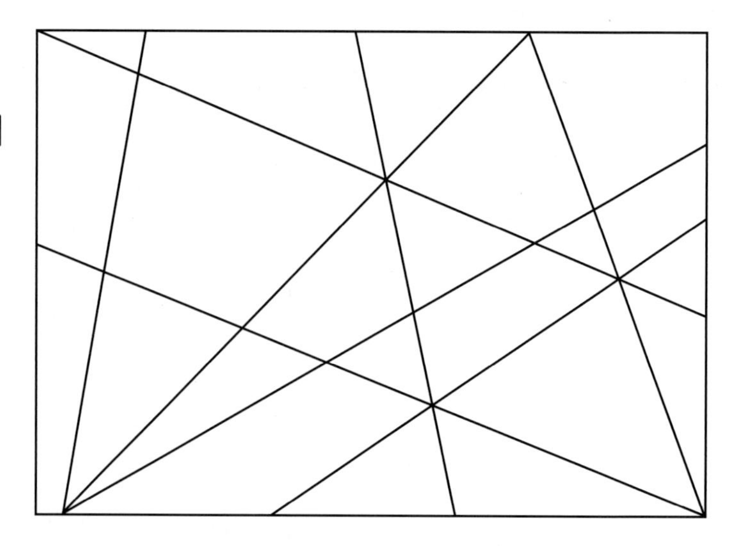

? What is the relationship between triangle and quadrilateral?

3 Let's draw a straight line in the quadrilateral and make the following shapes.

We can connect the vertices.

Akari

I can draw a straight line between the sides.

Yu

＼ Want to explore ／

? Purpose What shapes can you make by drawing a straight line?

❶ 2 triangles

❷ | triangle and | quadrilateral

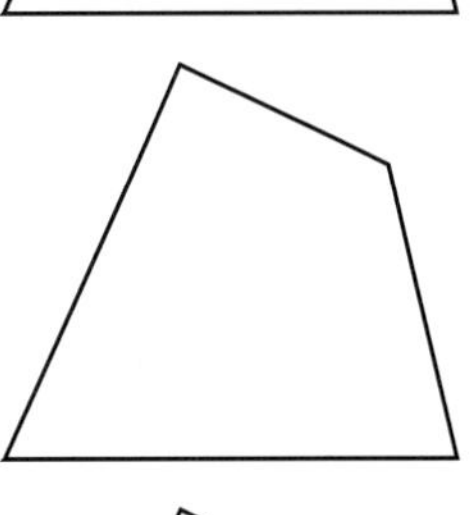

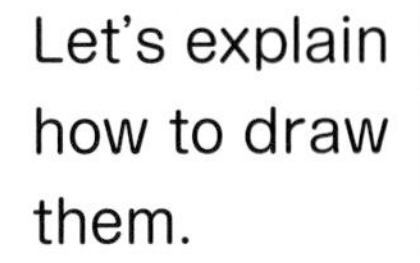
Let's explain how to draw them.

❸ 2 quadrilaterals

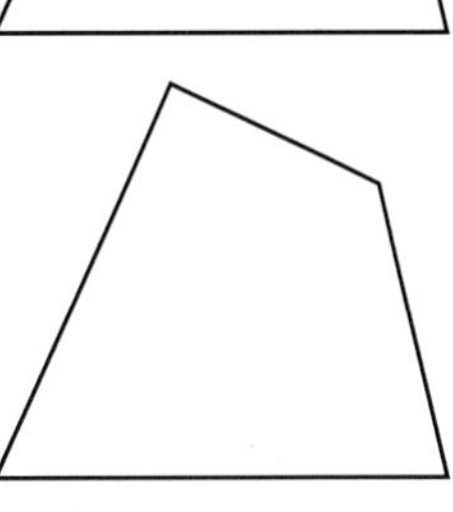

! Summary

If you draw | straight line in a quadrilateral, you can make two shapes such as triangles and quadrilaterals.

1 Let's draw a straight line in the triangle and make the following shapes.

① 2 triangles

② | triangle and | quadrilateral

2 Right angles

1 Let's fold a sheet of paper as shown below. 

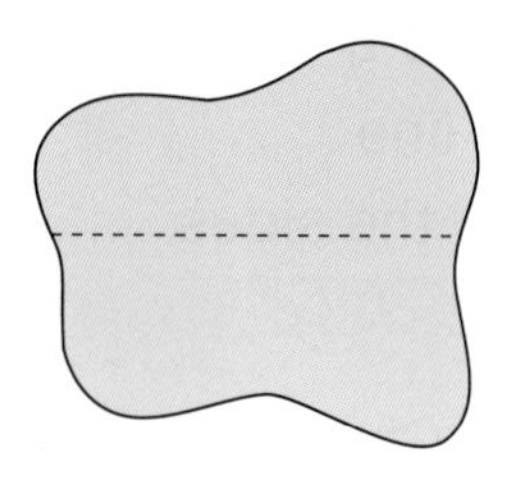

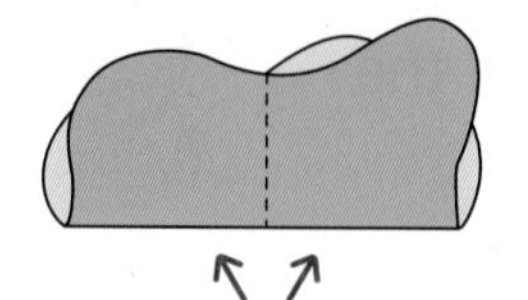

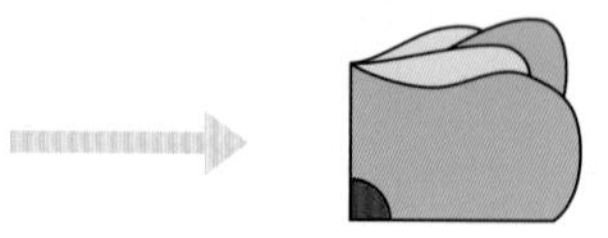

Fold the paper completely as shown.

Sara: It's the same as the corner of the notebooks and textbooks.

Want to explore

Purpose Will there be the same right angles using papers of any shapes?

Haruto

① Let's fold other papers to examine if there will be a right angle formed.

The corner that is formed by folding the paper as shown above is called a **right angle**.

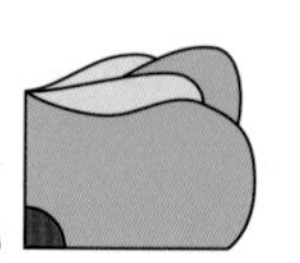

1 Is there a right angle in a triangle ruler? Let' s examine.

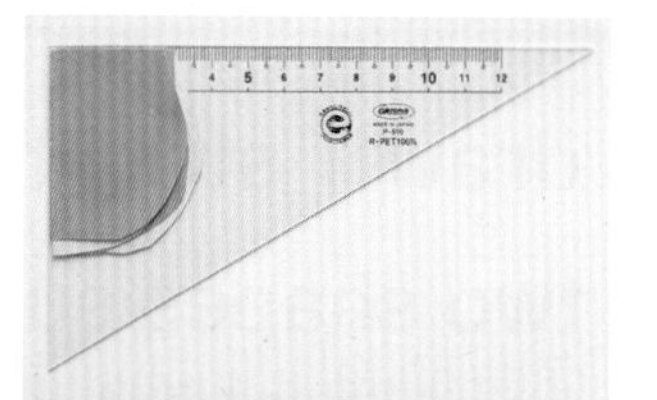

2 From your surroundings, let's look for right angles.

Akari: It seems that we can find a lot of right angles at school.

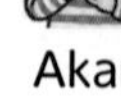

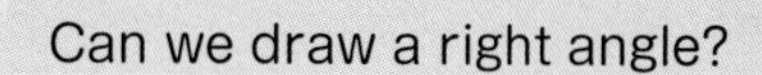

? Can we draw a right angle?

2 Let's draw many shapes with right angles by connecting the dots below.

Want to confirm

Purpose Let's check the right angles by using a triangle ruler.

Yu

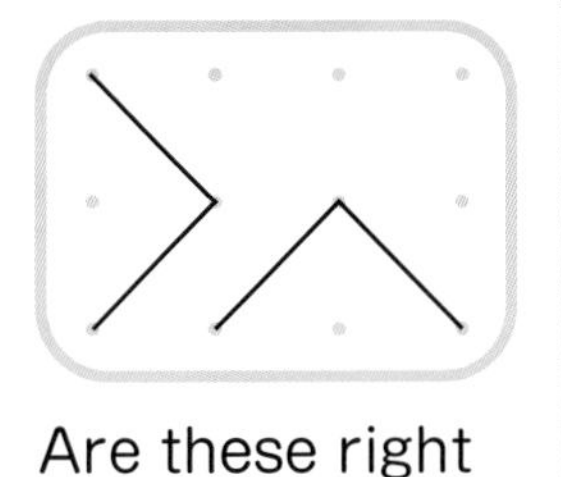

Are these right angles?

Akari

1 Let's draw a right angle using a triangle ruler. ▷

(1) Draw a horizontal straight line.

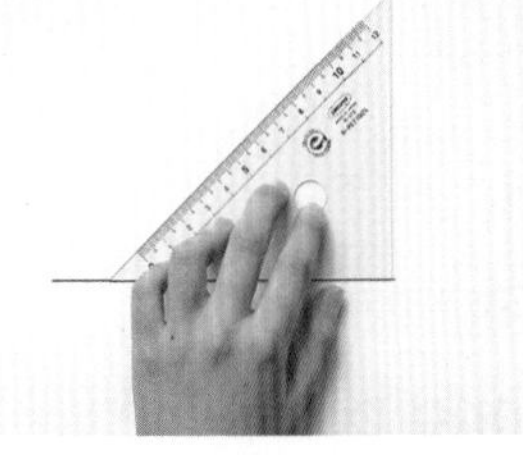

(2) Place an edge of the triangle ruler along the line you have just drawn.

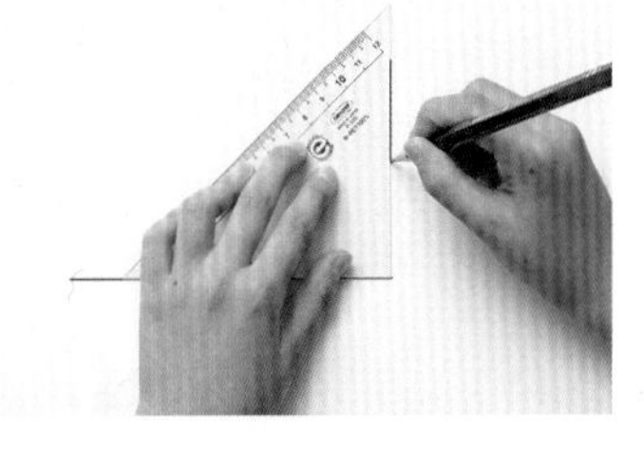

(3) Draw a vertical straight line.

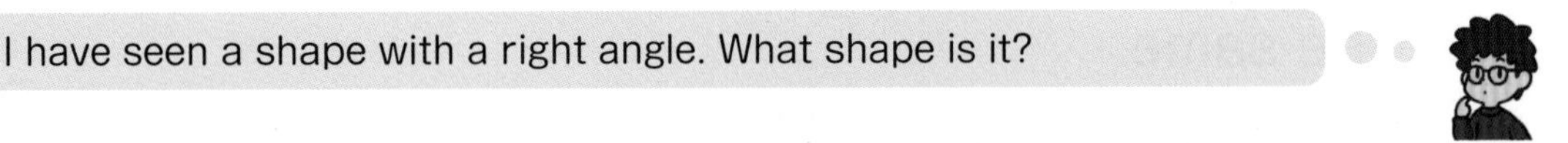

? I have seen a shape with a right angle. What shape is it?

3 Rectangles and squares

1 **Let's fold a sheet of paper as shown below and make right angles. What shape is formed?** ▷

The third right angle is formed at the same time as the fourth.

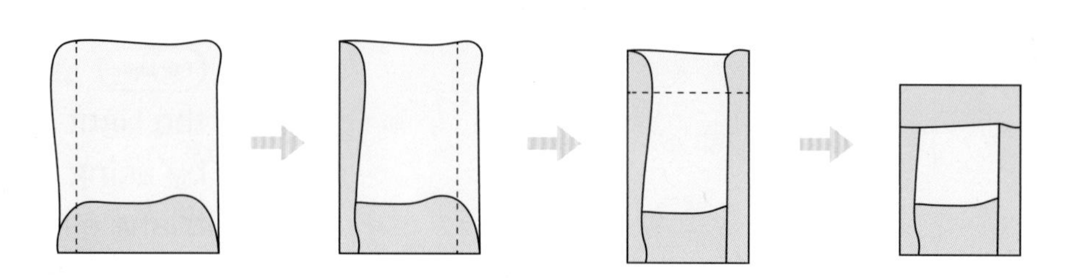

A quadrilateral where all 4 corners are right angles is called a **rectangle**.

\ Want to know /

What other properties do rectangles have?

It seems that there are sides with the same length.

1 Let's compare the lengths of the opposite sides of a rectangle.

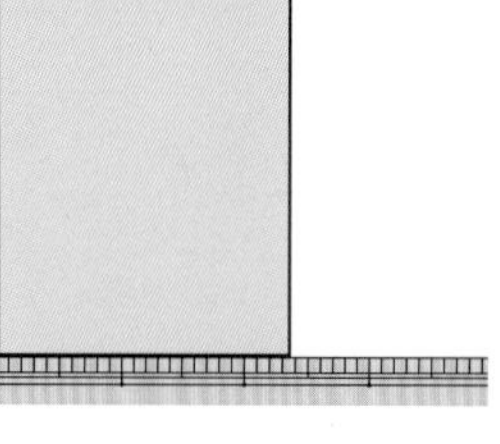

Measure and compare.

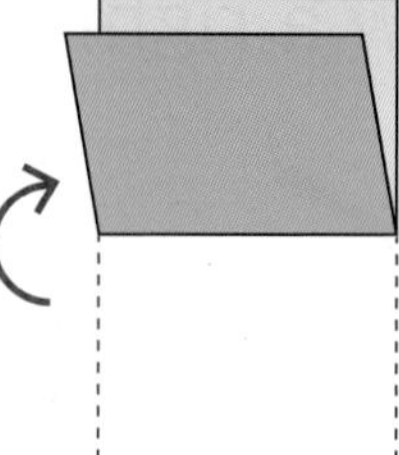

Fold and compare.

The lengths of the opposite sides of a rectangle are the same.

2 Which ones are rectangles?

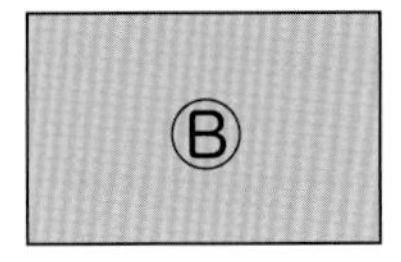

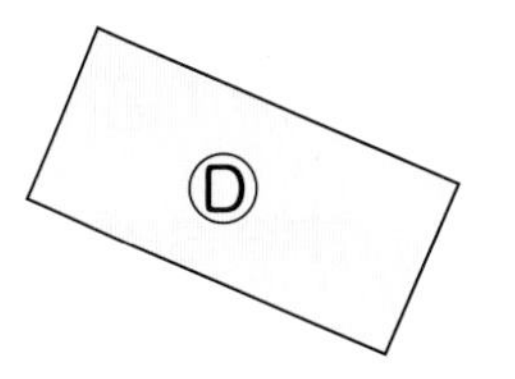

3 From your surroundings, let's look for things shaped rectangles.

What tool do we need to find a rectangle?

? Are there other shapes that has a right angle?

Squares ↓

2

Cut the rectangular paper as shown below. Open the folded paper. What shape is formed? ▷

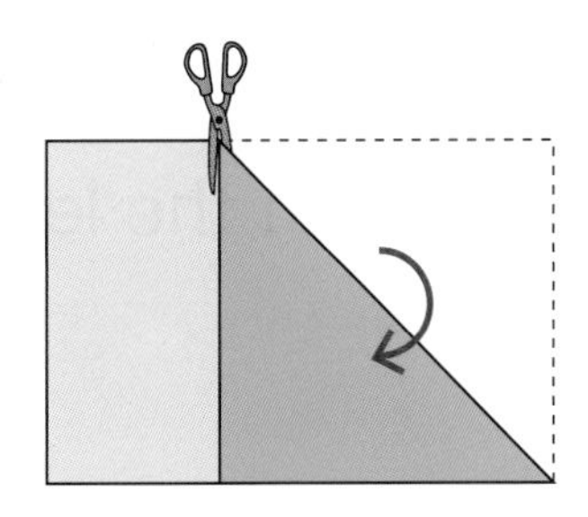

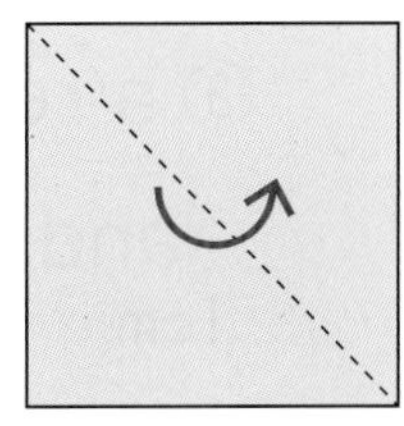

We get the same shape for the overlapped part.

Want to know

Purpose It looks similar to a rectangle, but what is the difference?

① Let's examine the four corners of the shape.

② Let's examine the four sides of the shape.

A quadrilateral where all 4 corners are right angles and all 4 sides are of the same length is called a **square**.

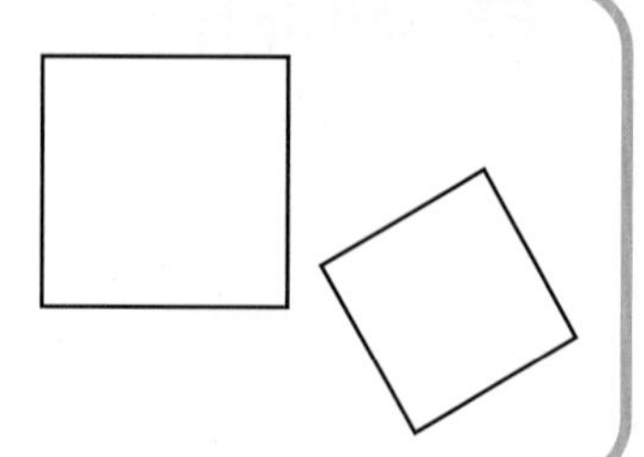

1 Which ones are squares?

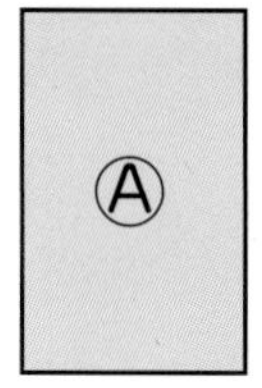

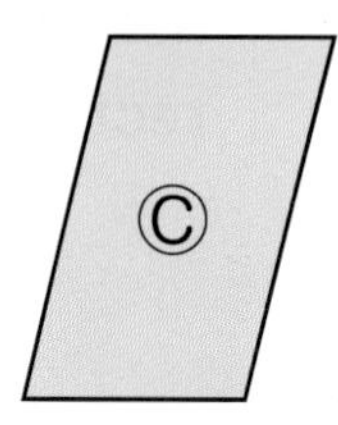

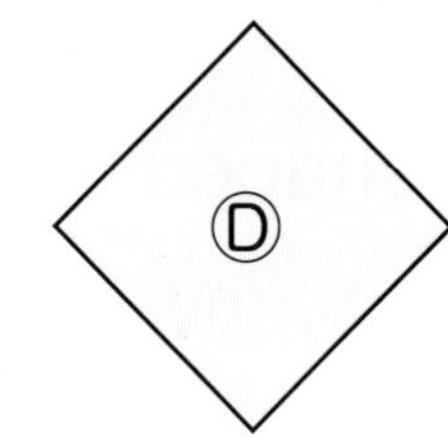

Summary The relationship between the lengths of the sides are different for squares and rectangles.

Haruto

2 From your surroundings, let's look for things shaped square.

Be sure that there are no overlaps.

3 Let's draw the following shapes.

① a rectangle of which the lengths of the sides are 3cm and 6cm

② a square of which the lengths of the sides are 4cm

③ a rectangle of which the lengths of the sides are 5cm and 4cm

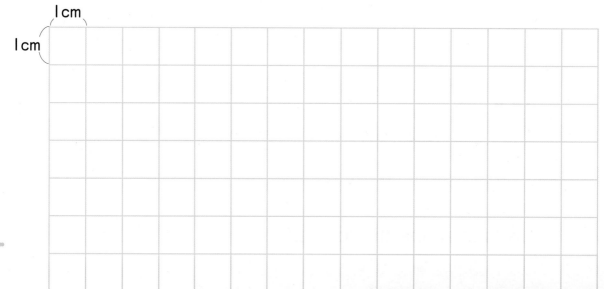

4 Right triangles

1 Let's cut the rectangular and square sheets of paper along the dashed lines as shown in the diagram.
Look at the shapes formed by cutting.

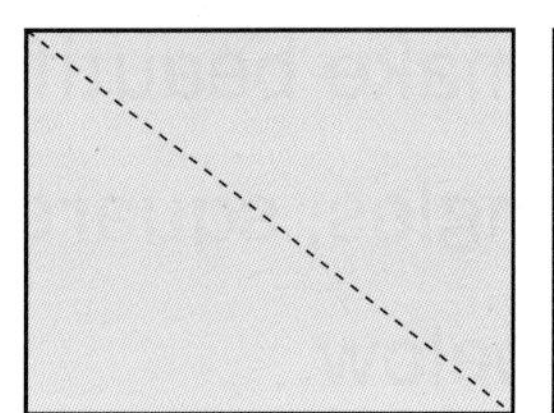

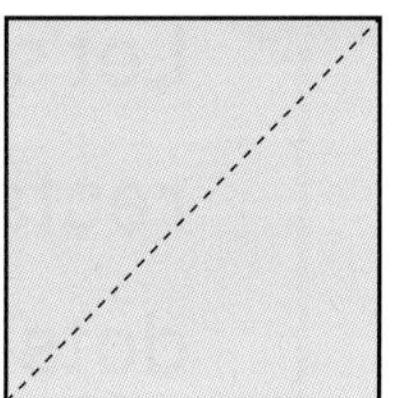

① Let's examine the shapes of the corners.

② What shapes are formed?

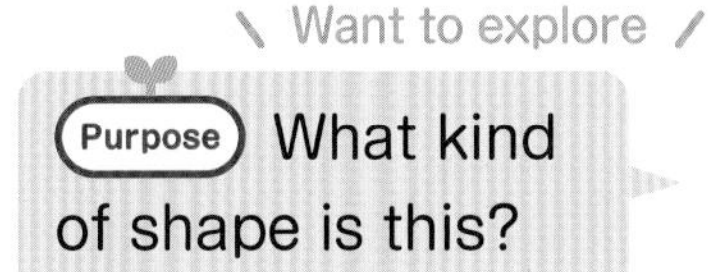

Want to explore
Purpose What kind of shape is this?

A triangle that has a right angle corner is called a **right triangle**.

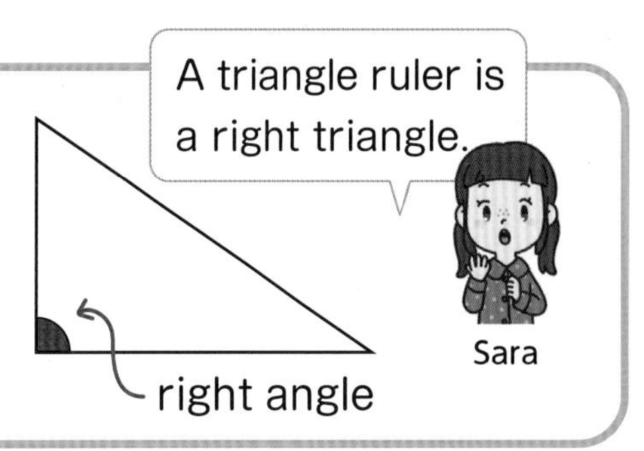

A triangle ruler is a right triangle.
right angle
Sara

1 Which ones are right triangles? Let's confirm by using a triangle ruler.

Can you guess before using a triangle ruler?

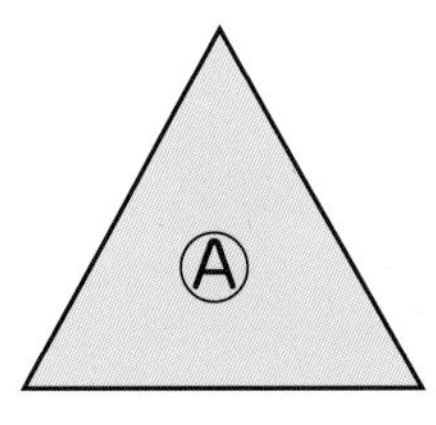

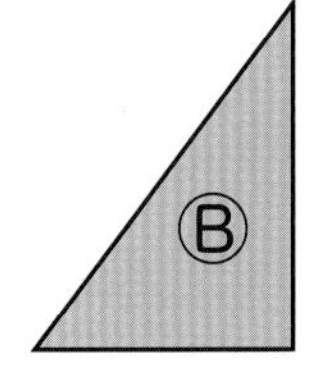

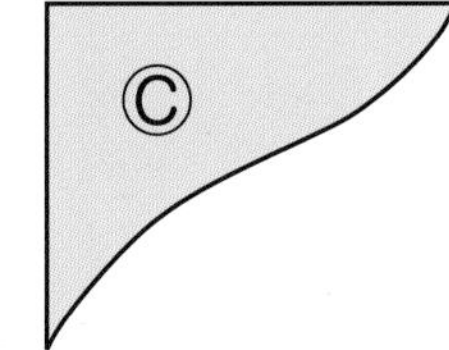

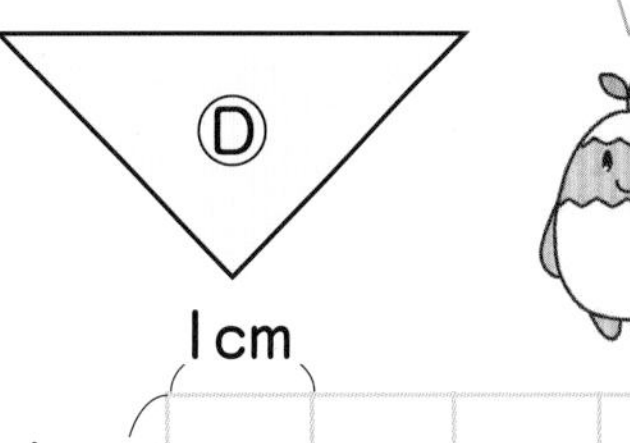

2 Let's draw a right triangle that has a right angle between the sides of 3cm and 4cm.

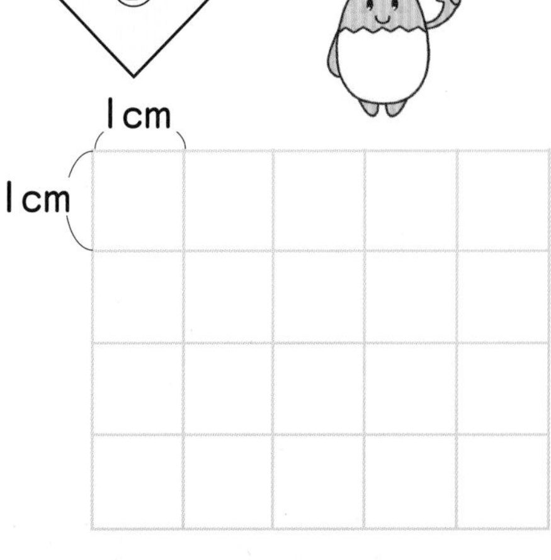

5 Making patterns

1 Let's make beautiful patterns by drawing rectangles, squares, or right triangles using the dots below.

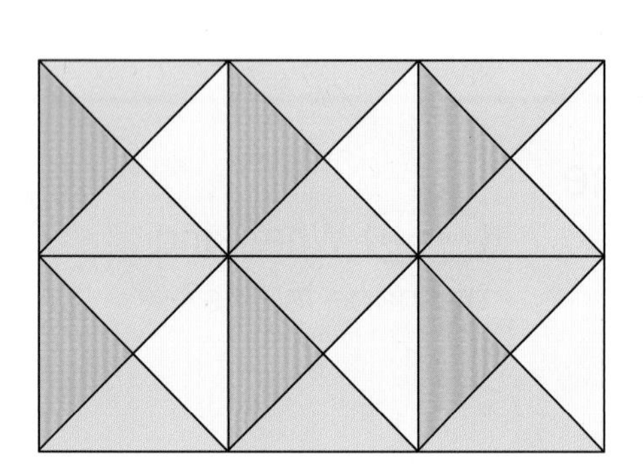

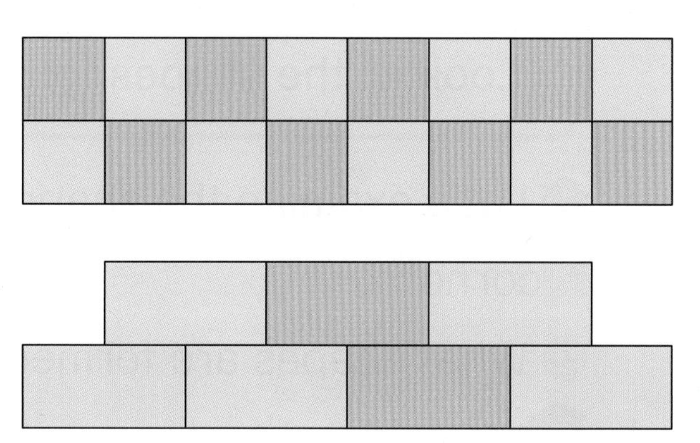

\ Want to try /

Can we make patterns by only using the same shapes?

Sara

CAN What can you do?

☐ We understand the properties of triangles and quadrilaterals. → pp.137 ~ 138

1 Let's fill in each ☐ with a number.

① There are ☐ sides and ☐ vertices in a triangle.

② There are ☐ sides and ☐ vertices in a quadrilateral.

☐ We understand the properties of rectangles and squares. → pp.142 ~ 145

2 Which of the following are rectangles, squares and right triangles?

Ⓐ Ⓑ Ⓒ Ⓓ Ⓔ Ⓕ Ⓖ

☐ We can draw a right triangle. → p.145

3 Draw a right triangle that has a right angle between the side of 2cm and the side of 3cm.

Supplementary Problems → p.158

Which "Way to See and Think Monsters" did you find in "11 Triangles and Quadrilaterals"?

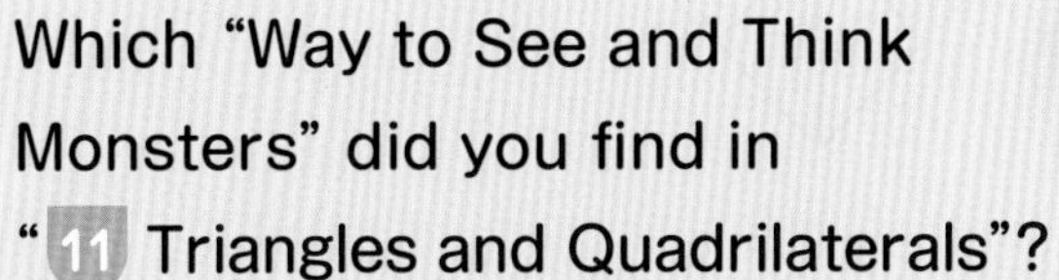

I found "Summarize" when I was categorizing the shapes according to the number of sides and their length.

What other monsters did you find?

Utilize Usefulness and Efficiency of Learning

1 Identify the following shapes.

① A quadrilateral in which all corners are right angles.

② A quadrilateral in which all corners are right angles and all sides have the same length.

③ A triangle with a right angle.

2 Let's draw a straight line in the rectangle and make the following shapes.

① 2 right triangles

② 2 squares

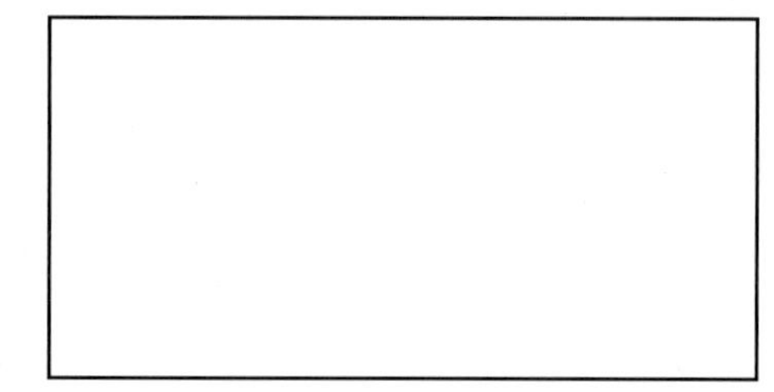

3 How many following shapes are there in the patterns?

① square

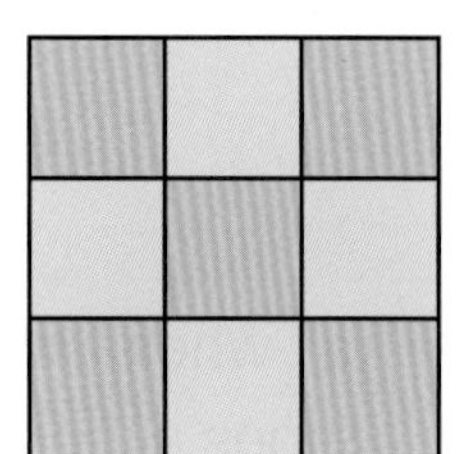

② right triangle

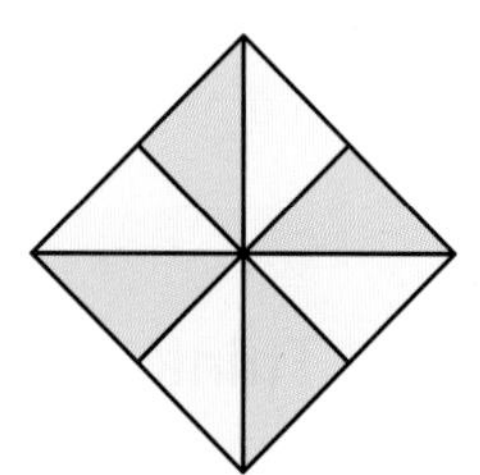

With the Way to see and think Monsters...

Let's Reflect!

Let's reflect on which monster you found in " 11 Triangles and Quadrilaterals."

Summarize

By categorizing the shapes according to their properties, we could summarize rectangle and square as quadrilateral, and right triangle as triangle.

① What is the property of rectangle, square, and right triangle?

rectangle

square

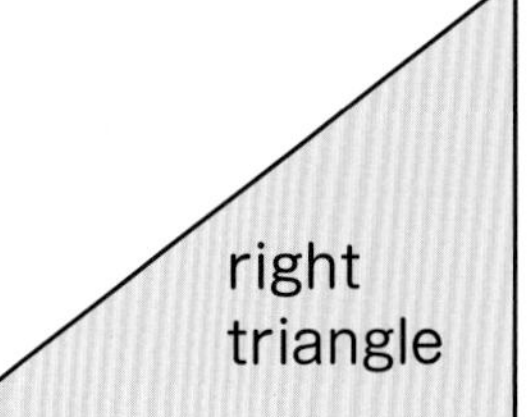

Rectangle is a shape where 4 corners are ☐ , and the length of the opposite sides are ☐ .

Square is a shape where all 4 corners are ☐ , and the length of all 4 sides are ☐ .

Right triangle is a shape that has a ☐ corner.

Haruto

Sara

Yu

? Solve the ?

There are various shapes in quadrilateral and triangle. We learned about rectangle, square, and right triangle.

Haruto

Want to Connect

Are there any other shapes that has a name?

Sara

More Math!

[Supplementary Problems]

[Let's deepen.]

[Answers]

1 Tables and Graphs

→ pp.12 ~ 19

1 The weather in March is shown in the table below.
Let's answer the following.

Weather in March

Weather	Sunny	Cloudy	Rainy	Snowy
Number of Days	12	9	8	2

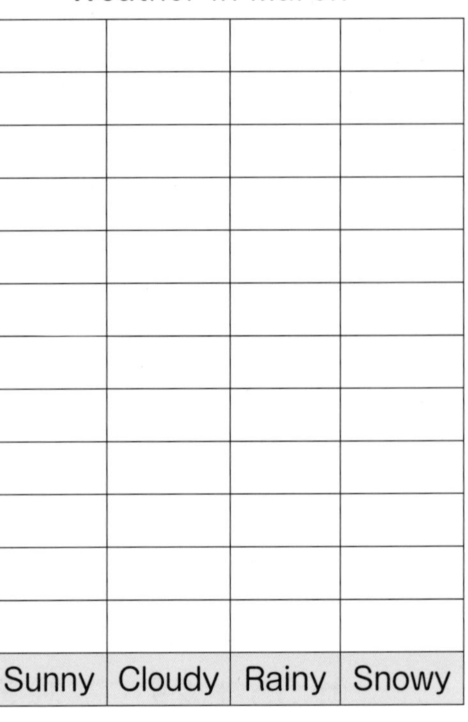

① Represent the number of days in each weather condition by using ○ on the graph.

② How many more sunny days are there than cloudy days?

2 Time and Duration (1)

→ pp.20 ~ 29

1 Let's answer the following.

① Express the time shown on clocks Ⓐ and Ⓑ.

② How long is the duration from time Ⓐ to time Ⓑ?

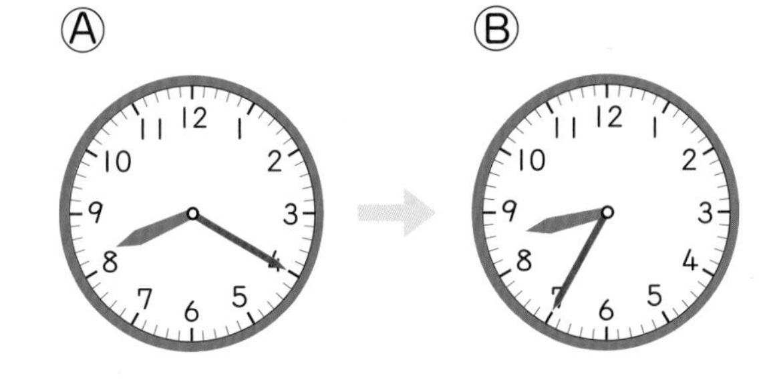

2 How long is the duration from time Ⓒ to time Ⓓ?

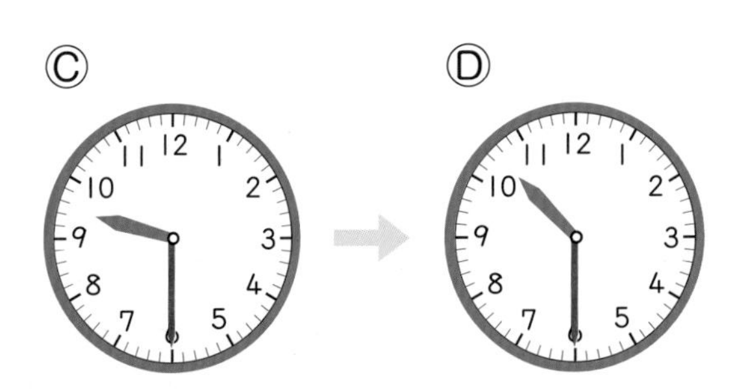

4 Addition in Vertical Form

→ pp.38 ~ 51

1 Let's calculate the following in vertical form.

① 23 + 34　② 29 + 40　③ 46 + 26　④ 58 + 37

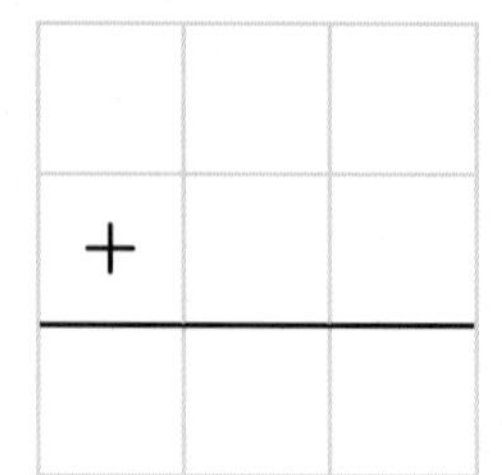

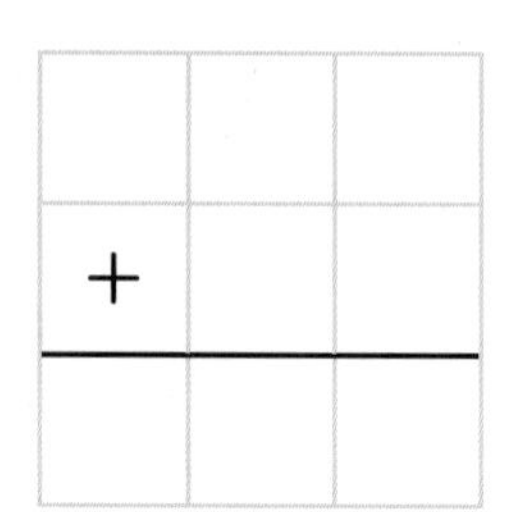

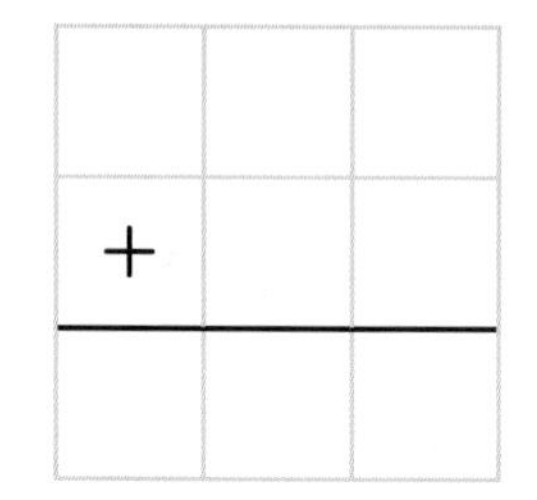

2 Let's calculate the following in vertical form.

① 42 + 36　② 26 + 33　③ 75 + 21　④ 36 + 20

⑤ 40 + 41　⑥ 10 + 80　⑦ 5 + 43　⑧ 8 + 31

⑨ 74 + 4　⑩ 52 + 7　⑪ 18 + 34　⑫ 47 + 35

⑬ 55 + 29　⑭ 38 + 13　⑮ 38 + 32　⑯ 74 + 6

⑰ 19 + 41　⑱ 63 + 9　⑲ 4 + 78　⑳ 3 + 57

3 There are 23 sheets of red paper and 5 sheets of blue paper. How many sheets are there in total?

4 Let's find the mistakes in the following processes and correct them.

①
```
  3 6
+ 2 8
-----
  5 4
```

②
```
  5 9
+ 4
-----
  9 9
```

③
```
  3 3
+   5
-----
  8 8
```

④
```
  6 3
+   7
-----
  6 0
```

5 The answer of the addition sentence below is 70. Let's fill in the □ with a number.

20 + □ = 70

5 Subtraction in Vertical Form

→ pp.52 ～ 63

1 Let's calculate the following in vertical form.

① 43 − 21 ② 59 − 8 ③ 60 − 23 ④ 52 − 47

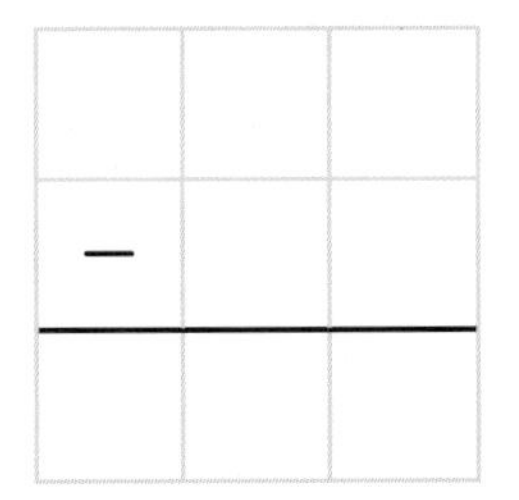

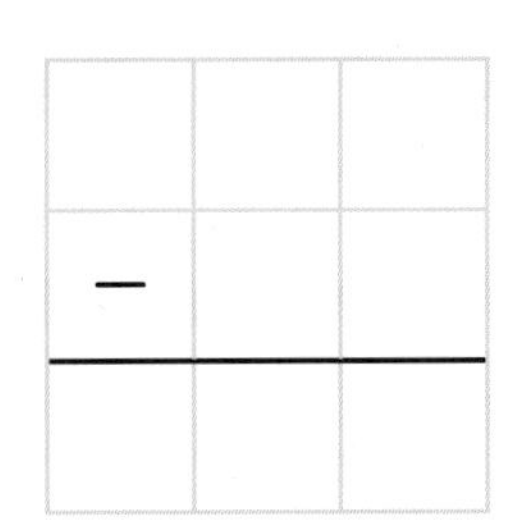

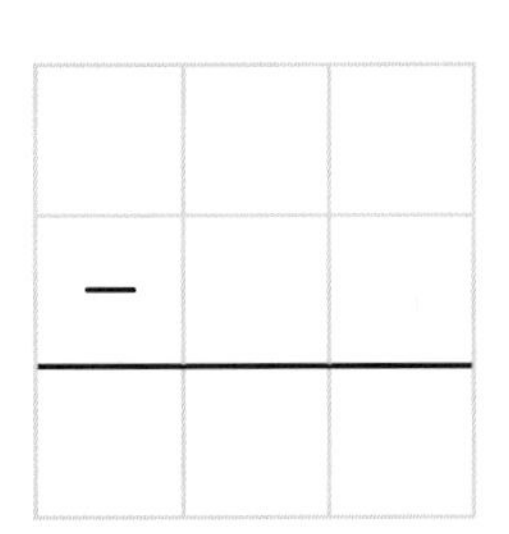

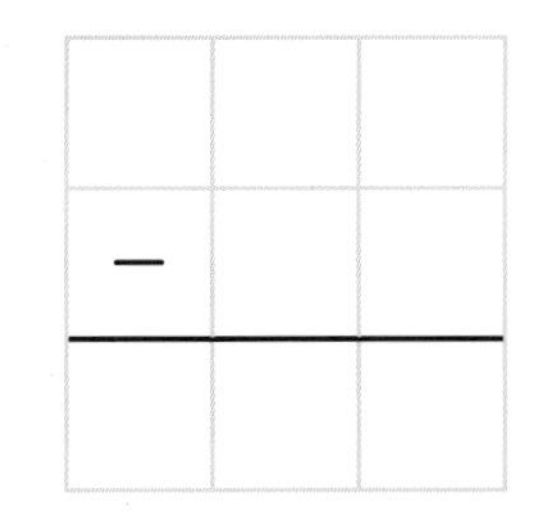

2 Let's calculate the following in vertical form.

① 87 − 52 ② 65 − 20 ③ 55 − 31 ④ 94 − 83
⑤ 49 − 3 ⑥ 56 − 26 ⑦ 38 − 31 ⑧ 74 − 4
⑨ 53 − 14 ⑩ 98 − 59 ⑪ 30 − 19 ⑫ 60 − 47
⑬ 44 − 36 ⑭ 72 − 63 ⑮ 34 − 5
⑯ 52 − 9 ⑰ 80 − 7 ⑱ 30 − 6

3 Let's calculate the following, and confirm the answer.

① 48 − 16 ② 52 − 29 ③ 27 − 9 ④ 90 − 8

4 Let's find the mistakes in the following processes and correct them.

① $\begin{array}{r} 63 \\ -\ 29 \\ \hline 44 \end{array}$ ② $\begin{array}{r} 78 \\ -\ 35 \\ \hline 33 \end{array}$ ③ $\begin{array}{r} 85 \\ -\ 4 \\ \hline 45 \end{array}$ ④ $\begin{array}{r} 60 \\ -\ 7 \\ \hline 63 \end{array}$

5 The answer of the subtraction sentence below is 20. Let's fill in the ☐ with a number.

50 − ☐ = 20

6 Length (1)

→ pp.64 ～ 78

1 Which tape is the longest?

Ⓐ Ⓑ Ⓒ

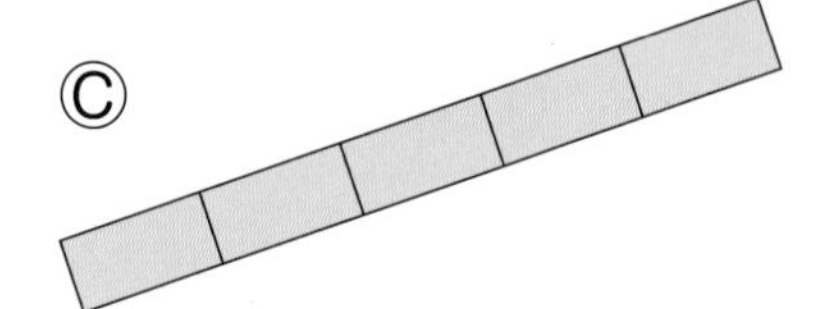

2 What are the lengths of the tapes and straight lines below in cm?

① ②

③ ④

3 What are the lengths of the tapes and straight lines below in cm and mm?

① ② ③

4 Let's draw a straight line for the given lengths below.

① 9cm ② 10cm5mm ③ 8cm7mm

5 Let's fill in each □ with a number.

① 7cm = □ mm ② 2cm8mm = □ mm

③ 80mm = □ cm ④ 93mm = □ cm □ mm

6 Which is longer, Ⓐ or Ⓑ?

① Ⓐ 4cm8mm Ⓑ 52mm

② Ⓐ 9cm Ⓑ 89mm

7 Let's calculate the following.

① 42cm + 36cm ② 15cm4mm + 9cm

③ 33cm − 18cm ④ 17cm5mm − 9cm

Addition and Subtraction (1)

→ pp.79 ~ 85

1 There are 28 boys and 26 girls.

How many children are there in total?

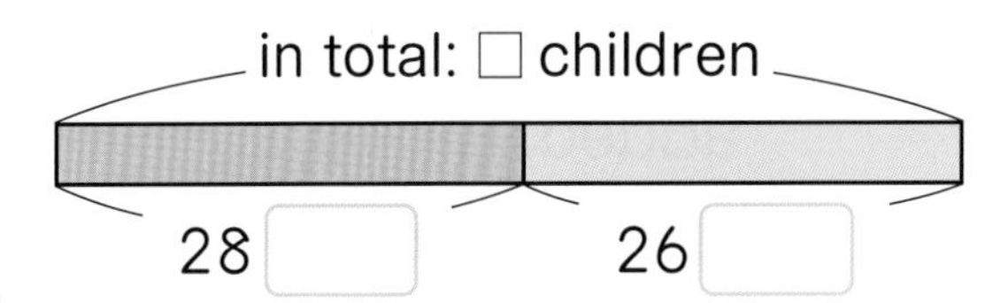

① Let's fill in the ☐ with a word.

② Let's write a math expression and the answer.

2 There were 40 cookies. I ate 18 of them. How many cookies are left?

① Let's fill in the ☐ with a word.

② Let's write a math expression and the answer.

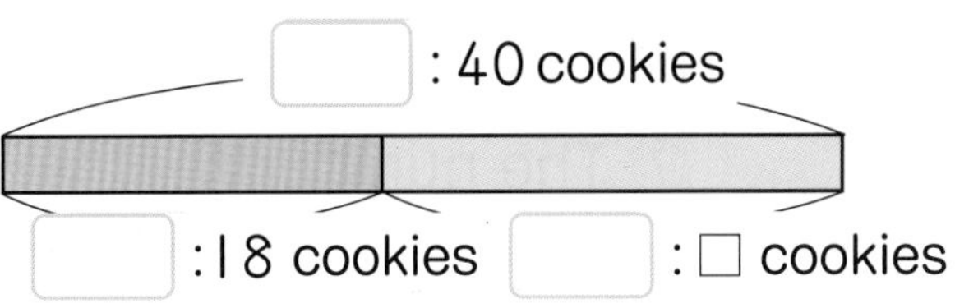

3 Yota has 34 sheets of colored paper. His brother has 8 more sheets than him. How many sheets does Yota's brother have?

① Let's fill in the ☐ with a number.

② Let's write a math expression and the answer.

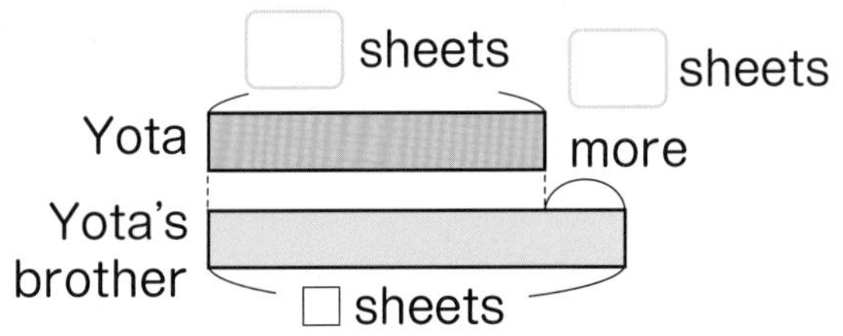

4 Akari has 35 marbles. Yuna has 7 fewer marbles than Akari. How many marbles does Yuna have?

① Let's fill in the ☐ with a number.

② Let's write a math expression and the answer.

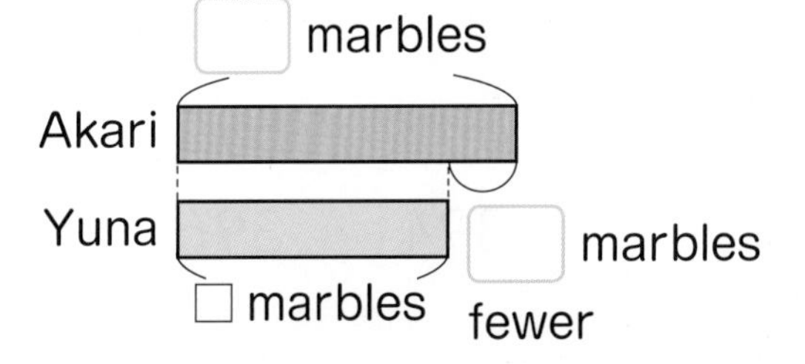

8 Numbers up to 1000

→ pp.86 ~ 99

1 Let's write the following numbers.

① The number when five hundred, thirty, and seven are added.

② The sum of 6 sets of 100, 5 sets of 10, and 4 ones.

③ The number that has 6 in the hundreds place, 0 in the tens place, and 2 in the ones place.

④ The number that is the sum of 7 sets of 100.

⑤ The number that is 200 smaller than 800.

⑥ The number that is 100 smaller than 1000.

⑦ The number that is 10 larger than 600.

2 Let's write the numbers that each ↑ is pointing.

3 Let's fill in each ☐ with a number.

① 270 is the sum of ☐ sets of 100 and ☐ sets of 10.

② 270 is the sum of ☐ sets of 10.

③ 600 is the sum of ☐ sets of 10.

It is also the sum of ☐ sets of 100.

④ The number that is the sum of 45 sets of 10 is ☐.

4 Let's fill in the ☐ with > or <.

① 685 ☐ 593　　② 776 ☐ 767

③ 394 ☐ 396　　④ 401 ☐ 410

Addition and Subtraction of Large Numbers

→ pp.100 ~ 117

1 Let's calculate the following in vertical form.

① 63 + 75 ② 85 + 41 ③ 52 + 92 ④ 69 + 60

⑤ 70 + 88 ⑥ 90 + 30 ⑦ 46 + 87 ⑧ 77 + 55

⑨ 95 + 26 ⑩ 62 + 68 ⑪ 63 + 39 ⑫ 87 + 16

⑬ 45 + 55 ⑭ 8 + 92 ⑮ 300 + 300

⑯ 500 + 200 ⑰ 400 + 70 ⑱ 600 + 50

⑲ 300 + 700 ⑳ 285 + 8 ㉑ 769 + 1

㉒ 145 + 26 ㉓ 328 + 54 ㉔ 846 + 35

2 Let's calculate the following in vertical form.

① 146 − 53 ② 159 − 85 ③ 119 − 72

④ 163 − 93 ⑤ 104 − 11 ⑥ 125 − 60

⑦ 147 − 68 ⑧ 173 − 95 ⑨ 182 − 86

⑩ 151 − 73 ⑪ 124 − 57 ⑫ 160 − 91

⑬ 104 − 38 ⑭ 106 − 19 ⑮ 100 − 84

⑯ 107 − 8 ⑰ 800 − 400 ⑱ 500 − 100

⑲ 1000 − 500 ⑳ 426 − 4 ㉑ 632 − 3

㉒ 114 − 8 ㉓ 345 − 36 ㉔ 260 − 53

3 Let's find the mistakes in the following processes and correct them.

①
$$\begin{array}{r} 106 \\ -\ \ 48 \\ \hline 68 \end{array}$$

②
$$\begin{array}{r} 453 \\ -\ 26\ \ \\ \hline 193 \end{array}$$

10 Amount of Water

→ pp.122 ~ 133

1 How many L and dL of water are there?

① 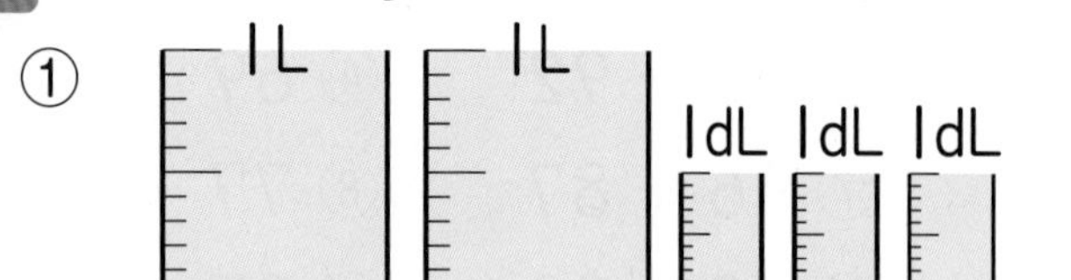

② 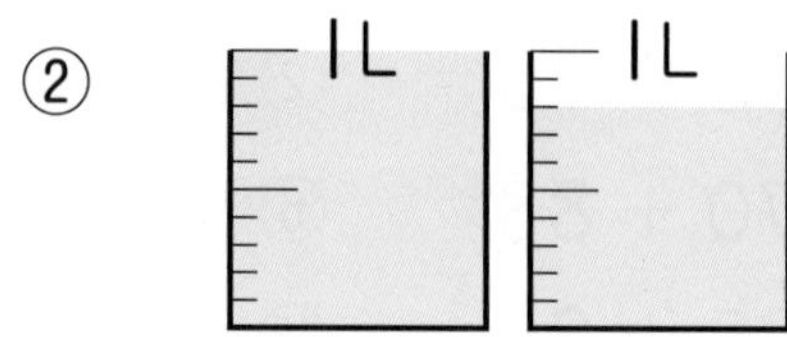

2 Let's fill in each ☐ with a number.

① 6L = ☐ dL
② 3L7dL = ☐ dL
③ 90dL = ☐ L
④ 42dL = ☐ L ☐ dL
⑤ 3dL = ☐ mL
⑥ 1000mL = ☐ L
⑦ 600mL = ☐ dL

3 Let's fill in the ☐ with >, <, or =.

① 3L4dL ☐ 34dL
② 1L8dL ☐ 20dL
③ 6L3dL ☐ 62dL

11 Triangles and Quadrilaterals

→ pp.134 ~ 149

1 Let's fill in each ☐ with a word or a number.

① 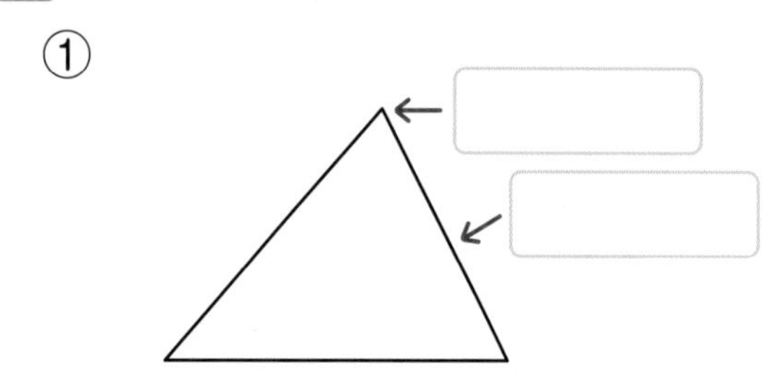

In a triangle, the number of sides is ☐ and the number of vertices is ☐.

② 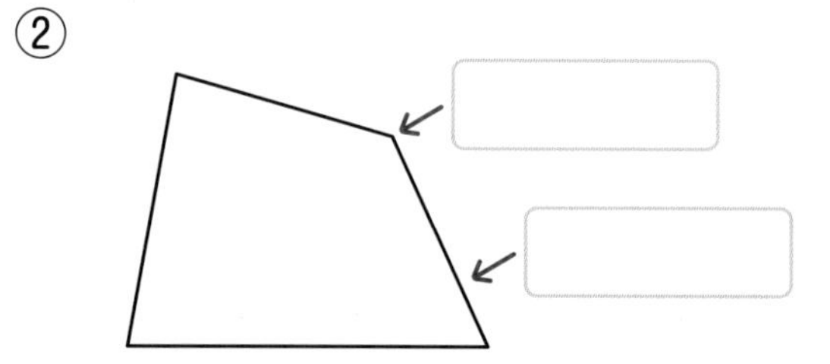

In a quadrilateral, the number of sides is ☐ and the number of vertices is ☐.

2 Which corner is a right angle?

①

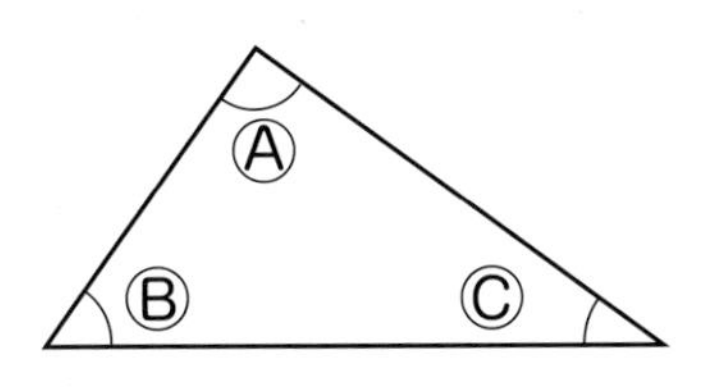

②

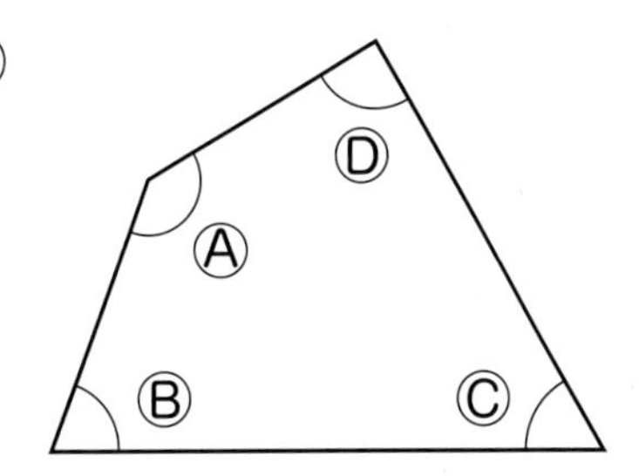

3 Look at the following rectangle and fill in each

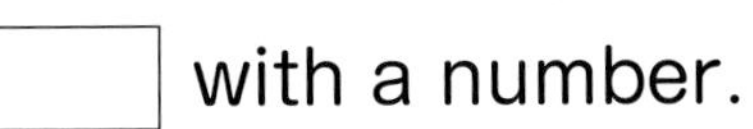

with a number.

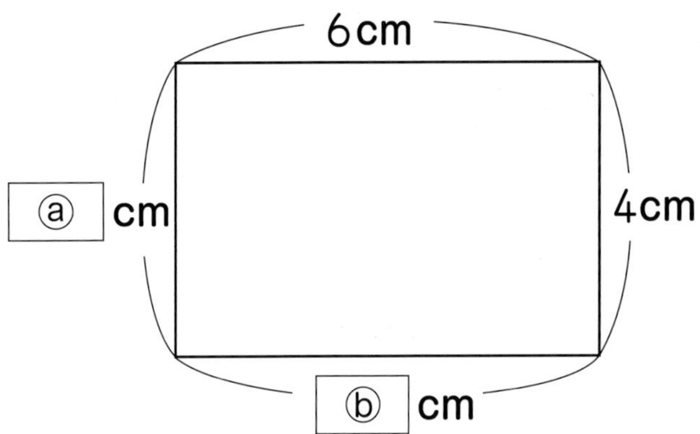

4 Let's draw the following shapes.

① A rectangle with sides of 3cm and 5cm.

② A square with a side of 4cm.

③ A right triangle.

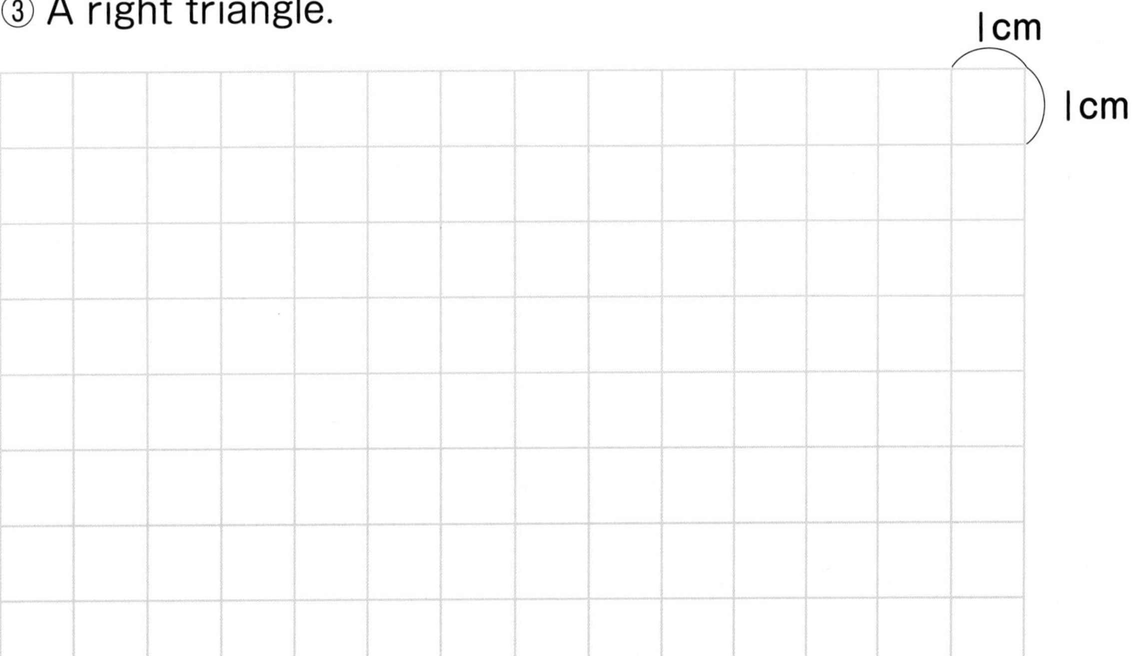

Let's make 100!

Numbers from 1 to 9 are arranged in ascending order like '1 2 3 4 5 6 7 8 9.' Let's make 100 that is the answer of a math expression by inserting + or − between the numbers.

If you insert neither + nor − between two numbers, the two numbers will be regarded as a 2-digit number.

① Let's make 100 by inserting + or − in the □.

(example)

1 + 2 + 3 − 4 + 5 + 6 + 78 + 9 = 100

1 □ 2 □ 3 □ 4 □ 5 □ 6 □ 7 □ 8 □ 9 = 100

1 □ 2 □ 3 □ 4 □ 5 □ 6 □ 7 □ 8 □ 9 = 100

1 □ 2 □ 3 □ 4 □ 5 □ 6 □ 7 □ 8 □ 9 = 100

1 □ 2 □ 3 □ 4 □ 5 □ 6 □ 7 □ 8 □ 9 = 100

② Numbers are arranged in descending order like '9 8 7 6 5 4 3 2 1.' Let's think about whether you can make 100.

Let's make 6 dL!

There is a 5dL cup, 8dL cup, and a big cup. How do you make 6dL by using these cups?

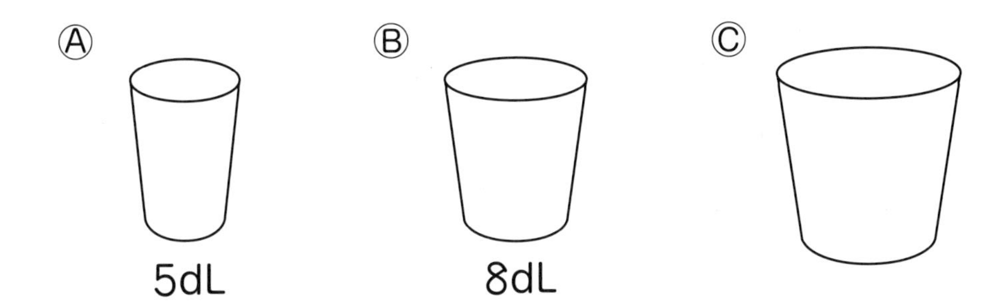

① Haruto has the following idea. Let's continue writing his idea.

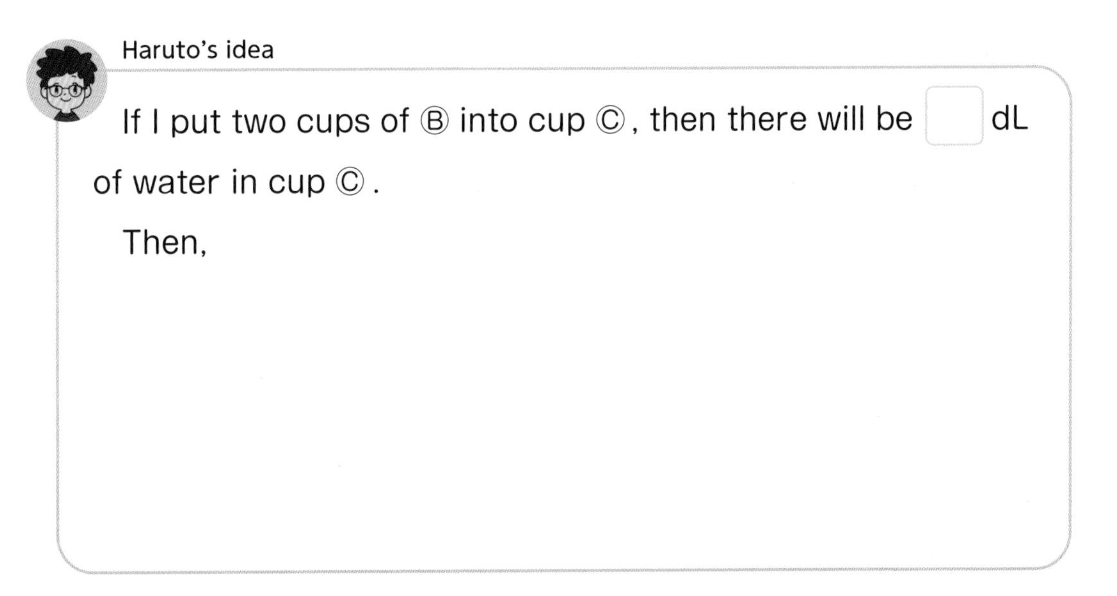

Haruto's idea
If I put two cups of Ⓑ into cup Ⓒ, then there will be ☐ dL of water in cup Ⓒ.
Then,

② Let's try to think of other ways.

Answers

[Supplementary Problems]

1 Tables and Graphs
→ p.151

1 ① Weather in March ② 3 days more

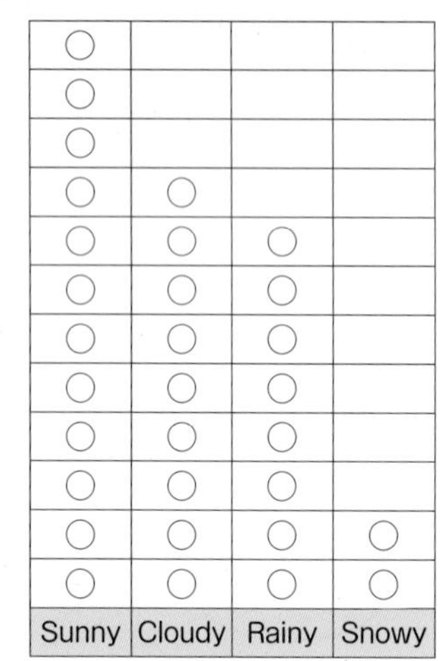

2 Time and Duration (1)
→ p.151

1 ① Ⓐ 8:20 Ⓑ 8:35
② 15 minutes

2 1hour

4 Addition in Vertical Form
→ p.152

1 ① 57 ② 69 ③ 72 ④ 95

2 ① 78 ② 59 ③ 96 ④ 56
⑤ 81 ⑥ 90 ⑦ 48 ⑧ 39
⑨ 78 ⑩ 59 ⑪ 52 ⑫ 82
⑬ 84 ⑭ 51 ⑮ 70 ⑯ 80
⑰ 60 ⑱ 72 ⑲ 82 ⑳ 60

3 28 sheets

4 ① $\begin{array}{r} 36 \\ +28 \\ \hline 64 \end{array}$ ② $\begin{array}{r} 59 \\ +\ \ 4 \\ \hline 63 \end{array}$ ③ $\begin{array}{r} 33 \\ +\ \ 5 \\ \hline 38 \end{array}$ ④ $\begin{array}{r} 63 \\ +\ \ 7 \\ \hline 70 \end{array}$

5 50

5 Subtraction in Vertical Form
→ p.153

1 ① 22 ② 51 ③ 37 ④ 5

2 ① 35 ② 45 ③ 24 ④ 11
⑤ 46 ⑥ 30 ⑦ 7 ⑧ 70
⑨ 39 ⑩ 39 ⑪ 11 ⑫ 13
⑬ 8 ⑭ 9 ⑮ 29 ⑯ 43
⑰ 73 ⑱ 24

3 ① Answer: 32 Confirmation of the answer: 32 + 16 = 48
② Answer: 23 Confirmation of the answer: 23 + 29 = 52
③ Answer: 18 Confirmation of the answer: 18 + 9 = 27
④ Answer: 82 Confirmation of the answer: 82 + 8 = 90

4 ① $\begin{array}{r} 63 \\ -29 \\ \hline 34 \end{array}$ ② $\begin{array}{r} 78 \\ -35 \\ \hline 43 \end{array}$ ③ $\begin{array}{r} 85 \\ -\ \ 4 \\ \hline 81 \end{array}$ ④ $\begin{array}{r} 60 \\ -\ \ 7 \\ \hline 53 \end{array}$

5 30

6 Length (1)
→ p.154

1 Ⓑ

2 ① 6cm ② 4cm ③ 5cm ④ 7cm

3 ① 2cm5mm ② 5cm8mm ③ 3cm8mm

4 (Omitted)

5 ① 70 ② 28 ③ 8 ④ 9, 3

6 ① Ⓑ ② Ⓐ

7 ① 78cm ② 24cm4mm
③ 15cm ④ 8cm5mm

7 Addition and Subtraction (1)
→ p.155

1 ①

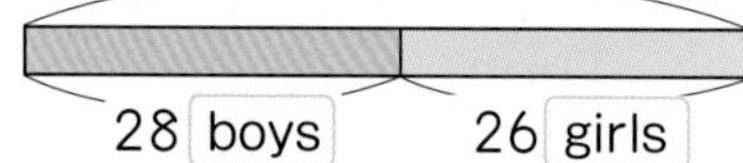

② 28 + 26 = 54 Answer: 54children

2 ①

cookies 40 cookies

eaten : 18 cookies left : □ cookies

② 40 − 18 = 22 Answer: 22 cookies

3 ①

34 sheets

Yota 8 sheets more

Yota's brother

□ sheets

② 34 + 8 = 42 Answer: 42 sheets

4 ①

35 marbles
Akari
Yuna
7 marbles fewer
□ marbles

② 35 − 7 = 28　　Answer: 28 marbles

8 Numbers up to 1000 → p.156

1 ① 537　② 654　③ 602　④ 700
⑤ 600　⑥ 900　⑦ 610

2 ① Ⓐ 173　Ⓑ 188
② Ⓒ 291　Ⓓ 306

3 ① 2, 7　② 27　③ 60, 6　④ 450

4 ① >　② >　③ <　④ <

9 Addition and Subtraction of Large Numbers → p.157

1 ① 138　② 126　③ 144　④ 129
⑤ 158　⑥ 120　⑦ 133　⑧ 132
⑨ 121　⑩ 130　⑪ 102　⑫ 103
⑬ 100　⑭ 100　⑮ 600　⑯ 700
⑰ 470　⑱ 650　⑲ 1000　⑳ 293
㉑ 770　㉒ 171　㉓ 382　㉔ 881

2 ① 93　② 74　③ 47　④ 70
⑤ 93　⑥ 65　⑦ 79　⑧ 78
⑨ 96　⑩ 78　⑪ 67　⑫ 69
⑬ 66　⑭ 87　⑮ 16　⑯ 99
⑰ 400　⑱ 400　⑲ 500　⑳ 422
㉑ 629　㉒ 106　㉓ 309　㉔ 207

3
① $\begin{array}{r} 106 \\ -\ 48 \\ \hline 58 \end{array}$
② $\begin{array}{r} 453 \\ -\ 26 \\ \hline 427 \end{array}$

10 Amount of Water → p.158

1 ① 2L3dL　② 1L8dL

2 ① 60　② 37　③ 9　④ 4, 2
⑤ 300　⑥ 1　⑦ 6

3 ① =　② <　③ >

11 Triangles and Quadrilaterals → pp.158 ~ 159

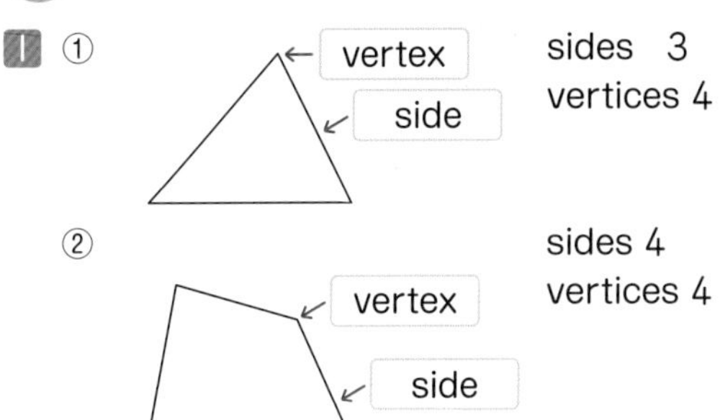

1 ① sides 3
vertices 4

② sides 4
vertices 4

2 ① Ⓐ　② Ⓓ

3 ⓐ 4　ⓑ 6

4

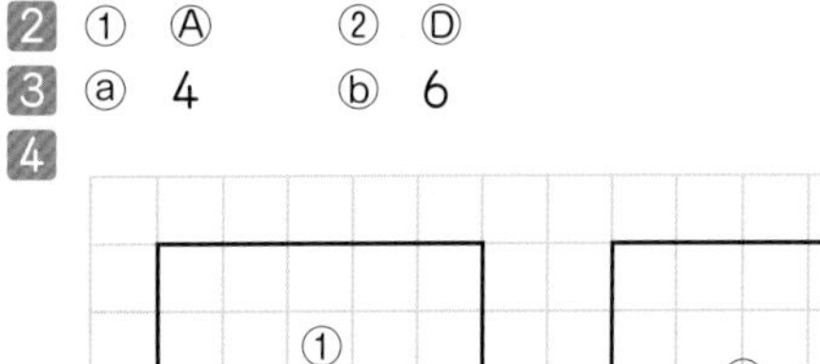

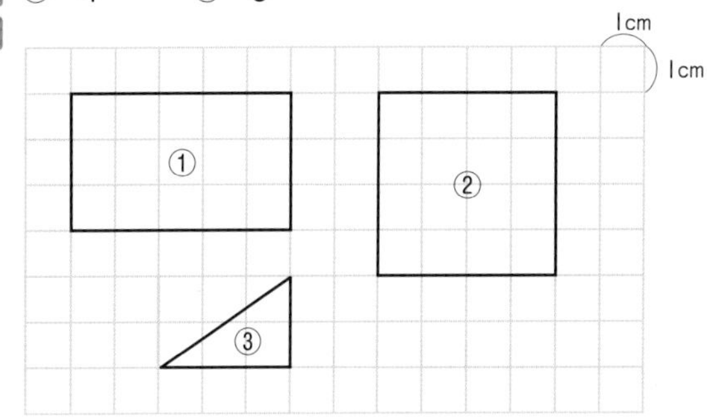

[Let's deepen.]

Let's make 100! → p.160

① (example)
123 − 45 − 67 + 89 = 100
123 + 45 − 67 + 8 − 9 = 100
12 − 3 − 4 + 5 − 6 + 7 + 89 = 100
1 + 2 + 34 − 5 + 67 − 8 + 9 = 100

② (example)
98 + 7 + 6 − 5 − 4 − 3 + 2 − 1 = 100

Let's make 6dL! → p.161

① we remove 2 cups of water in cup Ⓐ from cup Ⓒ. Then, 6dL will be left in cup Ⓒ.

② First, put water in cup Ⓑ and pour it into cup Ⓐ. There will be 3 dL left in cup Ⓑ. Put the 3dL in cup Ⓑ into cup Ⓒ. If we repeat this process one more time, there will be 6dL in cup Ⓒ.

words and Symbols

which we learned in this textbook

Tape

→ To be used in pp.64-67.
Please cut these out for use.

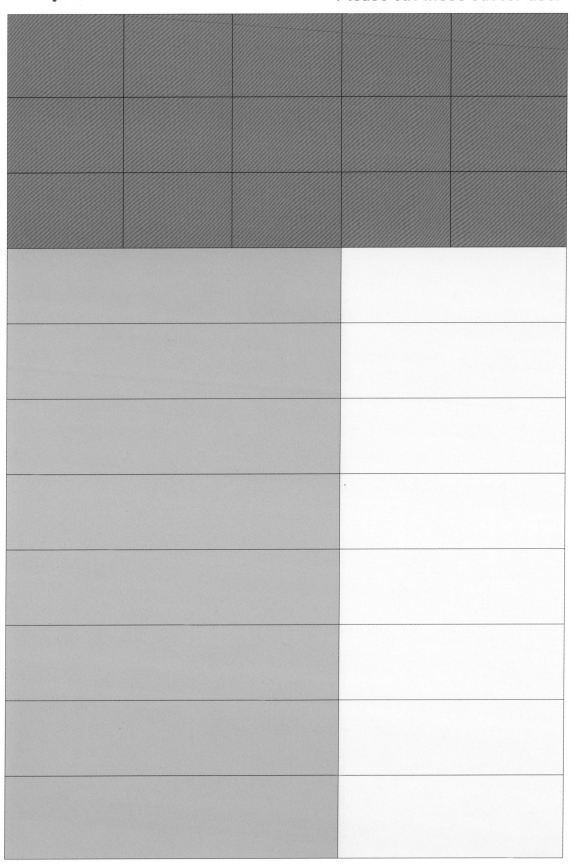

Memo

Editors of Original Japanese Edition

[Head of Editors]

Shin Hitotsumatsu (Kyoto University), Yoshio Okada (Hiroshima University)

[Supervising Editors]

Toshiyuki Akai (Hyogo University), Toshikazu Ikeda (Yokohama National University), Shunji Kurosawa (The former Rikkyo University), Hiroshi Tanaka (The former Tsukuba Univ. Elementary School), Kosho Masaki (The former Kokugakuin Tochigi Junior College), Yasushi Yanase (Tamagawa University)

[Editors]

Shoji Aoyama (Tsukuba Univ. Elementary School), Kengo Ishihama (Showa Gakuin Elementary School), Hiroshi Imazaki (Hiroshima Bunkyo University), Atsumi Ueda (Hiroshima University), Tetsuro Uemura (Kagoshima University), Yoshihiro Echigo (Tokyo Gakugei Univ. Setagaya Elementary School), Hisao Oikawa (Yamato University), Hironori Osawa (Yamagata University Graduate School), Tomoyoshi Owada (Shizuoka University), Nobuhiro Ozaki (Seikei Elementary School), Masahiko Ozaki (Kansai Univ. Elementary School), Kentaro Ono (Musashino University), Hiroshi Kazama (Fukui University), Michihiro Kawasaki (Oita University), Miho Kawasaki (Shizuoka University), Yoshiko Kambe (Tokai University), Yukio Kinoshita (Kwansei Gakuin Elementary School), Tomoko Kimura (Tamon Elementary School, Setagaya City), Satoshi Kusaka (Naruto University of Education), Kensuke Kubota (Naruo Higashi Elementary School, Nishinomiya City), Itsushi Kuramitsu (The former University of the Ryukyus), Maiko Kochi (Kounan Elementary School, Toshima City), Chihiro Kozuki (Daiyon. Elementary School, Hino City), Goto Manabu (Hakuoh University), Michihiro Goto (Tokyo Gakugei Univ. Oizumi Elementary School), Hidenori Kobayashi (Hiroshima Univ. Shinonome Elementary School), Akira Saito (Shibata Gakuen University Graduate School), Masahiko Sakamoto (The former Tokoha University Graduate School), Junichi Sato (Kunitachigakuen Elementary School), Hisatsugu Shimizu (Keio Yochisha Elementary School), Ryo Shoda (Seikei University), Masaaki Sugihara (University of the Sacred Heart, Tokyo), Jun Suzuki (Gakushuin Primary School), Shigeki Takazawa (Shiga University), Chitoshi Takeo (Nanzan Primary School), Hidemi Tanaka (Tsukuba Univ. Elementary School), Toshiyuki Nakata (Tsukuba Univ. Elementary School), Hirokazu Nagashima (Daisan. Elementary School, Kokubunji City), Minako Nagata (Futaba Primary School), Kiyoto Nagama (Hiyagon Elementary School, Okinawa City), Satoshi Natsusaka (Tsukuba Univ. Elementary School), Izumi Nishitani (Gunma University), Kazuhiko Nunokawa (Joetsu University of Education), Shunichi Nomura (Waseda University), Mantaro Higuchi (Kori Nevers Gakuin Elementary School), Satoshi Hirakawa (Showa Gakuin Elementary School), Kenta Maeda (Keio Yokohama Elementary School), Hiroyuki Masukawa (University of the Sacred Heart, Tokyo), Shoichiro Machida (Saitama University), Keiko Matsui (Hasuike Elementary School, Harima Town), Katsunori Matsuoka (Naragakuen University), Yasunari Matsuoka (Matsushima Elementary School, Naha City), Yoichi Matsuzawa (Joetsu University of Education), Satoshi Matsumura (Fuji Women's University), Takatoshi Matsumura (Tokoha University), Kentaro Maruyama (Yokohama National Univ. Kamakura Elementary School), Kazuhiko Miyagi (Homei Elementary School Affiliated with J.W.U), Aki Murata (University of California, Berkeley), Takafumi Morimoto (Tsukuba Univ. Elementary School), Yoshihiko Moriya (The former Kunitachigakuen Elementary School), Junichi Yamamoto (Oi Elementary School, Oi Town), Yoshikazu Yamamoto (Showa Gakuin Elementary School), Shinya Wada (Kagoshima University), Keiko Watanabe (Shiga University)

[Reviser]

Katsuto Enomoto (The former Harara Elementary School, Kagoshima City)

[Reviser of Special Needs Education and Universal Design]

Yoshihiro Tanaka (Teikyo Heisei University)

[Cover]

Photo : Tatsuya Tanaka (MINIATURE LIFE)

Design : Ai Aso (ADDIX)

[Text]

Design : Ai Aso, Takayuki Ikebe, Hanako Morisako, Miho Kikuma, Sawako Takahashi, Katsuya Imamura, Satoko Okutsu, Kazumi Sakaguchi, Risa Sakemoto (ADDIX), Ayaka Ikebe

[Illustrations]

Lico, Gami, Nanami Sumimoto, DOKOCHALUCHO, Ryoko Totsuka, Kinue Naganawa, Mayumi Nojima, Yoshie Hayashi, B-rise

[Photo・Observation data]

Showa Gakuin Elementary School, Ken Ogawa, DEPO LABO, Ritmo, Pixta, Kagoshima Prefectural Visitors Bureau, Okinawa Convention & Visitors Bureau